Lecture Notes in Artificial Intelligence (LNAI)

Other volumes of the Lecture Notes in Computer Science relevant to Artificial Intelligence:

Lecture Notes in Artificial Intelligence

Subseries of Lecture Notes in Computer Science
Edited by J. Siekmann

Lecture Notes in Computer Science

Edited by G. Goos and J. Hartmanis

Editorial

Artificial Intelligence has become a major discipline under the roof of Computer Science. This is also reflected by a growing number of titles devoted to this fast developing field to be published in our Lecture Notes in Computer Science. To make these volumes immediately visible we have decided to distinguish them by a special cover as Lecture Notes in Artificial Intelligence, constituting a subseries of the Lecture Notes in Computer Science. This subseries is edited by an Editorial Board of experts from all areas of AI, chaired by Jörg Siekmann, who are looking forward to consider further AI monographs and proceedings of high scientific quality for publication.

We hope that the constitution of this subseries will be well accepted by the audience of the Lecture Notes in Computer Science, and we feel confident that the subseries will be recognized as an outstanding opportunity for publication by authors and editors of the AI community.

Editors and publisher

Lecture Notes in Artificial Intelligence

Edited by J. Siekmann

Subseries of Lecture Notes in Computer Science

459

R. Studer (Ed.)

Natural Language and Logic

International Scientific Symposium
Hamburg, FRG, May 9–11, 1989
Proceedings

 Springer-Verlag

Berlin Heidelberg New York London
Paris Tokyo Hong Kong Barcelona

Editor

Rudi Studer
Institut für Angewandte Informatik und
Formale Beschreibungsverfahren
Universität Karlsruhe
Kaiserstraße 12, D-7500 Karlsruhe 1, FRG

CR Subject Classification (1987): I.2.4, I.2.7

ISBN 3-540-53082-7 Springer-Verlag Berlin Heidelberg New York
ISBN 0-387-53082-7 Springer-Verlag New York Berlin Heidelberg

This work is subject to copyright. All rights are reserved, whether the whole or part of the material
is concerned, specifically the rights of translation, reprinting, re-use of illustrations, recitation,
broadcasting, reproduction on microfilms or in other ways, and storage in data banks. Duplication
of this publication or parts thereof is only permitted under the provisions of the German Copyright
Law of September 9, 1965, in its current version, and a copyright fee must always be paid.
Violations fall under the prosecution act of the German Copyright Law.
© Springer-Verlag Berlin Heidelberg 1990
Printed in Germany

Printing and binding: Druckhaus Beltz, Hemsbach/Bergstr.
2145/3140-543210 – Printed on acid-free paper

Foreword

This volume contains the papers presented at the International Scientific Symposium "Natural Language and Logic", which was held in Hamburg, May 9 - 11, 1989. The symposium was one in a series of symposia which IBM Germany organizes at regular intervals.

The choice of the topic was motivated by recent developments in the field of computational linguistics and knowledge representation. In both areas, the application as well as the need for further development of logic-based methods have gained in importance. Moreover, the development of attributive concept description languages within the framework of unification-based grammars, on the one hand, and within the framework of order-sorted logics and KL-ONE like knowledge representation formalisms, on the other hand, has led to a rather close collaboration between researchers from computational linguistics and artificial intelligence. Since such a collaboration is also a main characteristic of the LILOG project (LInguistic and LOGic Methods for Understanding German Texts) at the Scientific Center of IBM Germany, the topic for the Hamburg symposium "Natural Language and Logic" evolved quite naturally.

I would like to thank all the people who have helped to make the Hamburg symposium a success. Special thanks are due to P. Bosch, O. Herzog and C.-R. Rollinger, who have provided valuable support in setting up the symposium's programme, and to I. Leister and H. Statz, without whom we would not have had such a smooth organization in such a pleasant environment.

Finally, I would like to express my thanks to G. Schillinger for having provided excellent support in handling all the editorial activities.

Karlsruhe, July 1990   Rudi Studer

Table of Contents

Treatment of Anaphoric Problems In Referentially Opaque Contexts

Arendse Bernth
IBM Research Division
T.J. Watson Research Center
P.O. Box 704
Yorktown Heights, NY 10598

Abstract

This paper describes the treatment of some anaphoric expressions in certain referentially opaque contexts by the experimental discourse understanding system **LODUS**. Noun phrases occurring in potentially opaque contexts are given one of two possible interpretations: (1) The noun phrase describes the way the agent refers to the entity in question, or (2) the noun phrase acts solely as a means of identifying the referent, regardless of the linguistic expression used for referring to it. The first case, which is the opaque case, is handled by complex variables called roles. Here a further distinction is made between value-loaded and value-free roles, indicating whether the expression is referential or not. The second, and transparent, case is handled like other non-opaque cases. Some examples of the treatment in **LODUS** of opaque contexts are given, among these Quine's Ralph and Ortcutt story.

1 Introduction

This paper describes the treatment of some anaphoric expressions in certain referentially opaque contexts by **LODUS** (Logic-Oriented Discourse Understanding System) [Bernth 1988, 1989]. **LODUS** is an experimental discourse understanding system set in a logic programming framework, which uses inference to resolve references and to maintain consistency. The underlying theory, **CDS** (Computational Discourse Semantics) [Bernth 1988, 1989], draws inspiration from Discourse Representation Theory [Kamp 1981] and Situation Semantics [Barwise and Perry 1983], but is designed for use by computational systems and thus focuses on how intelligent behavior of such systems may be achieved by inference on the context. **LODUS** does not have a parser of its own, but uses a **Slot Grammar** parser [McCord 1989, 1990], which produces **LFL** (Logical Form Language) expressions [McCord 1981, 1987] for single sentences of a discourse. Each **LFL** expression is interpreted in the context consisting of the discourse processed so far, together with domain knowledge, and then embedded into the discourse structure, thus creating *one* CDS structure out of several **LFL** expressions. This interpretation makes use of an inference system based on an intuitionistic theorem prover by [Haridi and Sahlin 1988]. All parts of this system including the parser and the theorem prover are written in Prolog. The system is currently running on an IBM 3090.

The paper falls into two parts: The first part contains a general discussion of some of the problems encountered in handling anaphora in referentially opaque contexts, whereas the second part describes the actual treatment by **LODUS**.

2 Anaphoric Expressions In Opaque Contexts

In this section, we discuss some of the problems encountered in handling anaphoric expressions in referentially opaque contexts. For this purpose, we start out by giving examples of what is meant by anaphoric expressions and opaque contexts. In this context we also look at the distinction between the so-called *de dicto* and *de re* interpretations of noun phrases. Following this, we discuss how the de re and de dicto interpretations may be combined with attributive and referential uses of noun phrases.

2.1 Anaphoric Expressions

Anaphoric expressions include, but are not limited to, the following:
- Noun phrases with definite articles
- Pronouns
- Proper names
- Temporal references

Including proper names in this list may be controversial, but reflects the general treatment of proper names in **CDS** and **LODUS**, where the *first* occurrence of a proper name is considered non-anaphoric, but *subsequent* uses of a name are considered anaphoric; they have as antecedents entities introduced earlier in the discourse and are treated like other anaphoric expressions.[1] In this paper, however, our main concern is noun phrases with definite determiners and third person pronouns. In fact, temporal references are only mentioned for the sake of completeness and will not be discussed further.

For the purposes of this paper, we take *resolution of anaphoric expressions* to mean the identification of the correct antecedent. This antecedent is sought in the context consisting of previous discourse, world knowledge, and assumptions. *Assumptions* consist of whatever assumptions and attitudes a given person might have; other terms for this concept are *belief worlds* and *derived context*.

---

[1]   A further discussion of the treatment of proper names may be found in [Bernth 1988].

2.2 Referential Opacity

Referential opacity covers the phenomenon that expressions denoting the same object are not mutually substitutable *salva veritate* in certain contexts. Furthermore, it may be the case that the expression does not have a referent in the real world – hence *existential import* fails. Referential opacity may occur in cases where somebody, whom we choose to refer to as the *agent*, has a certain mental relation to a proposition, hence the common name for verbs denoting such relations: *propositional attitude verbs*. Examples are **think, believe, doubt**, etc. Also some verbs expressing certain relations to *individuals* may introduce referential opacity, e.g. **seek**.

Consider the example:

Oedipus thinks that Jocasta is childless, but she isn't.

Jocasta is Oedipus's mother, so under normal circumstances we should be able to substitute this for **Jocasta**:

Oedipus thinks that his mother is childless, but she isn't.

Of course this does not make sense if we take the embedded noun phrase (**his mother**) to describe the way Oedipus actually refers to this person. For this reason we shall distinguish between the so-called *de dicto* and *de re* interpretations.[2] The literature shows a slight confusion on the meaning of these terms, but we shall interpret them as follows:

De dicto: The way the agent refers is important and is described by the embedded noun phrase which cannot be substituted for another – this gives opacity.

De re: The embedded noun phrase is used solely as identification of some object without claiming that this is the way the agent refers to it – this gives transparency.

Other examples of opacity-creating constructions include:

- Past and future tenses: **Gorbachev will meet the President of U.S.**
- Certain sentential adverbs: **Necessarily, the Morning Star is the Evening Star.**
- Certain adjectives: **Former president.**

Another way of describing the distinction between de dicto and de re interpretations is in terms of belief worlds. The description given by a de dicto interpretation becomes part of the agent's belief world, whereas the description given by a de re interpretation of non-anaphoric

---

[2] Let it be pointed out here that – for all our references to Situation Semantics – we do not subscribe to the strict Realism of this theory. That would make the idea of de dicto interpretations meaningless.

expressions goes into the outer context representing the "real world" or the maximal belief world (the belief world on "God's" level).

2.3 Occurrences of Anaphoric Expressions

Anaphoric expressions may appear in opaque contexts as well as in transparent contexts. Sometimes the anaphoric expressions refer *out* of the opaque context:

1. John bought a car and Harry believes that the car broke down.
2. John bought a car and Harry believes that it broke down.
3. Bill wears a kilt. John believes he is a woman.

In all of these examples, an anaphoric expression appears in a potentially opaque context and refers out of this context to an entity introduced in a transparent context. In the first example, the anaphoric expression **the car** may or may not describe the way Harry refers to the car John bought. In the second example, the anaphoric expression consists of a third person singular pronoun. This raises the question: does this kind of pronoun contain enough of a description that a de dicto interpretation is possible? Personally, I find this hard to accept, at least in the cases where the pronoun refers *out* of the opaque context, and it is in accordance with [Evans 1977] that only a de re, i.e. a transparent, interpretation is possible. It should be mentioned, however, that [Richards 1981] disagrees with this. A similar point can be made about the third example. Here the antecedent of **he** is ambiguous between **Bill** and **John**, but this does not influence the question of opacity.

Anaphoric expressions may also refer *into* the opaque context or *within* it:

1. John dreams about a unicorn. The unicorn has a golden mane.
2. John dreams about a unicorn. It has a golden mane.
3. John tries to catch a fish and eat it.
4. John tries to catch a fish and wants to fry it for dinner.

The two first examples show how an opacity-creating construction may scope over more than one sentence. Exactly because of this, it is possible that a third person pronoun *can* be given a de dicto interpretation in certain cases (the second example). Here the opaque context created by **dreams about** scopes over the occurrence of **it** in the following sentence. This sentence should just be taken as an elaboration of the description of how John refers to the unicorn he is dreaming about. Hence, giving a third person pronoun a de dicto interpretation does not mean that the pronoun describes how the agent refers to the entity in question; rather it should be taken to mean that the reference should be resolved within the agent's belief world and whatever is predicated of that entity should go into his belief world.

The two last examples show how an opaque context can scope over an anaphoric expression occurring within the same sentence (example three) and how the introduction of an entity into an agent's set of assumptions makes it possible for that entity to be referred to in another opaque context (the last example).

Generally, these examples illustrate two points: first, it is possible for a noun phrase occurring in an opaque context to be the antecedent of an anaphoric expression that textually appears in a seemingly transparent context. Second, the last example shows that it is possible for such a noun phrase to be the antecedent of an anaphoric expression that textually appears in another opaque context. Furthermore, the agent of the attitude verb embedding the anaphoric expression does not have to be the same as the agent of the attitude embedding the antecedent, as illustrated by Geach's example:

    Hob believes that a witch has blighted his mare, and Nob believes
    that the witch has bewitched his cow.

As mentioned earlier, the context includes assumptions or belief worlds, and we may add here that even though each person has a different set of assumptions, parts of these assumptions will be shared with other persons. In this example, Hob and Nob share some assumptions about a witch. This does not appear directly from the discourse, but it is the only way the discourse makes sense.

Generally, it is probably preferable to find an antecedent in a transparent context rather than in an opaque context, other things being equal. This corresponds to the usual linguistic principle of preferring a simpler construction over a more complex one. But the examples shown above warrant the addition of the following principles: If an antecedent for an anaphoric expression appearing in a seemingly transparent context cannot be found in a transparent context, then look for the antecedent in an opaque context. And: If an antecedent for an anaphoric expression appearing in an embedded clause (opaque position) cannot be found within the agent's set of assumptions (including the real world), then look for the antecedent in someone else's set of assumptions.

2.4 Attributive/Referential Interpretations

Definite descriptions appearing in transparent contexts have been explained as having two different uses by [Donnellan 1966], viz. the attributive and the referential uses.[3] These can be combined with the de re and de dicto interpretations in interesting ways.

The attributive use states something about whoever is the so-and-so:

    Smith's murderer [whoever it is] is insane.

whereas the referential use picks out a certain referent:

    Smith's murderer [viz. Mr. Jones] is insane.

The important characteristic of the referential use is that it picks out an *individual*; the attributive use gives a *description*, which some individual might satisfy. [Barwise and Perry 1983] explains this in terms of *value-loaded* and *value-free* interpretations. The value-free interpretation of a definite description is a relation, a partial function from situations to individuals; the value-loaded is the value of that function when applied to some particular context. The value-free interpretation hence gives a *description*, and the value-loaded an *individual*. Regarding value-free (i.e. attributive) interpretations, we shall restrict our attention here to occurrences in referentially opaque contexts, since they pose the most interesting problems. In referentially transparent contexts we shall assume value-free interpretations. That is, regarding attributive uses, we shall only consider what Situation Semantics calls *inner* attributive interpretations, which are the value-free interpretations creating an opaque context. Furthermore, we shall regard the distinction between attributive and referential uses of definite noun phrases as a parallel to the distinction between *specific* and *non-specific* uses of indefinite descriptions. This means that we shall use the term "attributive" to cover "non-specific" and "referential" to cover "specific".

Whereas the distinction between the referential and attributive uses is a matter of *value-loaded* vs. *value-free* interpretations, the de re/de dicto distinction is a matter of *scope*, i.e. whether the quantification introduced by the noun phrase in the embedded sentence scopes over the attitude verb or vice versa. The referential/attributive uses and the de dicto/de re interpretations are thus different kinds of distinctions and combinations of them are possible. They can be combined in the following ways:

1. **Referential-de re.** A de re interpretation will always combine with a referential interpretation, since there has to be some specific object for the de re interpretation:

---

[3]   Probably, Donnellan only had in mind using these terms for describing the speaker's intentions, as indicated by the word *use*, but of course a similar distinction can be made for the addressee's interpretation of an utterance. Taking the speaker's intentions into consideration poses very interesting problems; in this paper, however, we shall limit our attention to the *addressee's* interpretation.

John believes that a (particular) president is powerful.

(The combination of an attributive and a de re interpretation does not really make sense, since the de re interpretation requires that there be a particular thing that the attitude is about, and the attributive requires that we do not point to a particular thing).

2. **Referential-de dicto.** There is a specific object which the agent refers to in a certain way:

    Charlie Brown believes that the little red-haired girl is beautiful.

    There is a specific person Charlie Brown is thinking of, and he does not know her name, but refers to her as **the little red-haired girl**.

3. **Attributive-de dicto.** There is a non-specific object the agent refers to in a certain way:

    John believes that the man who stole his bicycle (whoever it is) is red-haired.

    Here John does not know who stole his bicycle, but he believes that whoever it is, he must be red-haired (maybe because he left some red hairs behind).

3 Opacity in CDS and LODUS

As mentioned above, **LODUS** takes as input logical forms for single sentences, successively embedding them into one **CDS** structure. As in Situation Semantics, a **CDS** structure consists of a number of *situations*. Unlike Situation Semantics, however, *any* situation is described by two main parts: first a situation *type* and then a list of *anchors*, indicating how the variables of the situation type are bound to so-called reference markers to give the desired *realization* (instance) of the situation type. The situation type is furthermore divided into a *location* consisting of space and time coordinates, and a set of *conditions*. Each condition takes the form of a list, the head of which is the predicate, followed by its arguments. The division into a situation type and anchorings makes it easy to refer to a *type* of situation as well as a specific instance. The specific instance is created by *anchoring* the situation type to the reference markers specified in the list of anchors. This is easily done by unification of logical variables.

In order to handle opacity created by de dicto interpretations, a special kind of variable called a *role* is used. Roles are complex variables **R1 in Sit** consisting of a simple variable **R1** and a constraining situation type **Sit** which has **R1** as a constituent. The constraining situation type describes the agent's way of referring.

As explained earlier, the attributive de dicto interpretation and the referential de dicto interpretation have in common that the *way* the agent refers to the object in question is important. They differ, however, in their reference. In the attributive de dicto interpretation, the reference is value-free or non-specific. In the referential de dicto interpretation, the reference is value-loaded or specific. The value-free (attributive de dicto) interpretation is obtained by not anchoring the role variable R1, i.e. we use an unrealized situation type to describe the referent. The referential de dicto interpretation is obtained by using the role in a *value-loaded* way. This is done by anchoring the role variable *within the describing situation*, i.e. we use a situation to describe the referent.

As an example consider the possible interpretations of the sentence:

John believes that the winner is happy.

The attributive de dicto interpretation of this (*John believes that the winner (whoever it may be) is happy*) may be represented (partly) as

...[believes,X,[happy, R1 in [Loc,[winner,R1],anchor([ ])]]]...

where X is anchored to the reference marker for **John**. In this example the role variable **R1** is unanchored, which by way of the unrealized situation type describing **R1** gives us an attributive interpretation of **the winner**, meaning that John has the conviction that the person who wins − whoever it may be − must be happy.

If, on the other hand, John has a specific person in mind (maybe he sees the smiling winner on the stand) − this is the referential de dicto interpretation: *John believes of somebody that he refers to as "the winner" that he is happy* − we can represent this by anchoring **R1** to the appropriate reference marker:

...[believes,X,[happy, R1 in [Loc,[winner,R1],anchor([R1 to person1])]]]...

where **person1** is a reference marker that may be introduced into the discourse structure somewhere outside the role. In this case the situation type describing how John refers to the person in question is realized by a reference to the particular person John has in mind, thus giving us a referential de dicto interpretation. This illustrates the difference between roles and λ-expressions: the internal anchoring of **R1** still preserves the role and thus does not correspond to λ-conversion.

Finally, we have the referential de re interpretation (*John believes of the winner that he is happy*):

...[winner,X2] ... & [believe,X1,[happy,X2]]...

In this case, only the fact that John attributes happiness to the entity identified by **the winner** is added to his belief world, without any explicit statement about how John actually refers to it.

Let us now look at an example involving the verb **seek**:

1. John seeks a girl. He finds her.

2. John seeks a girl. He finds one.

Both discourses start out "**John seeks a girl**", which can be given both a referential de re interpretation and an attributive de dicto interpretation. It is not until the following sentence that the choice is clear. So the choice depends on the context.[4] If the following sentence is "**He finds her**", the referential de re interpretation makes good sense. Given a referential de re interpretation, the **CDS** structure for the discourse "**John seeks a girl. He finds her.**" is the following:

```
CDS structure:

    loc(X1,X2)
       [John,X3] &
       [girl,X4] &
       [seek,X3,X4]
       anchor
           X1 to t1
           X2 to s1
           X3 to john1
           X4 to girl1
&
    loc(t1,s1) overlap loc(now,here)
&
    loc(X5,X6)
       [find,X7,X8]
       anchor
           X5 to t2
           X6 to s2
           X7 to john1
           X8 to girl1
&
    loc(t2,s2) overlap loc(t1,s1)

57 ms.
```

In this example, **a girl** introduces the condition [**girl,X4**] and the anchoring of X4 to **girl1** into the discourse structure on the same level as [**John,X3**] and [**seek,X3,X4**]. The variable indicating the object of **find**, viz. **X8**, is anchored directly to the reference marker for the girl.

---

[4]   At the present stage, the LODUS implementation of CDS does not automatically choose the correct kind of interpretation on the basis of the context. The desired kind of interpretation is set by switches operated by the user.

If, on the other hand, the following sentence is "**He finds one**", we are likely to interpret the second sentence as "He finds one object satisfying the description given in the previous sentence", and this forces an attributive de dicto interpretation of the first sentence. The representation of **one** in the second sentence is then given as a realization of the situation type used in the first sentence to describe the kind of object John is looking for. This interpretation gives the following **CDS** structure for the discourse "**John seeks a girl. He finds one.**":

```
CDS structure:

    loc(X1,X2)
       [John,X3] &
       [seek,X3,X4 in
           loc(X5,X6)
              [girl,X4]
              anchor
       ]
       anchor
           X1 to t1
           X2 to s1
           X3 to john1
&
    loc(t1,s1) overlap loc(now,here)
&
    loc(X7,X8)
       [find,X9,X10 in [loc(X11,X12),[girl,X10],anchor([X10 to girl1])]]
       anchor
           X7 to t2
           X8 to s2
           X9 to john1
&
    loc(t2,s2) overlap loc(t1,s1)

78 ms.
```

This example shows the difference between a value-loaded interpretation (the referential interpretation), where John seeks a particular girl, and the value-free interpretation (the attributive de dicto interpretation), where John seeks someone who qualifies as a girl. The value-free interpretation makes use of a *role* describing the condition(s) that the individual John is seeking must fulfill. In the second sentence this role is value-loaded by the anchoring of the role variable to the reference marker **girl1** indicating the realization of the situation type.

The following example shows the **CDS** structure created by **LODUS** for one of the previously given examples:

John dreams about a unicorn. It has a golden mane.

Here **dreams about** can be taken to create an opaque context for **a unicorn**:

CDS structure:

```
loc(X1,X2)
   [John,X3] &
   [dream_about,X3,X4 in
       loc(X5,X6)
          [unicorn,X4]
          anchor
   ]
   anchor
       X1 to t1
       X2 to s1
       X3 to john1
&
   loc(t1,s1) overlap loc(now,here)
```

6 ms.

The next sentence continues John's dream, and should hence become part of the role describing the unicorn:

CDS structure:

```
loc(X1,X2)
   [John,X3] &
   [dream_about,X3,X4 in
       loc(X5,X6)
          [unicorn,X4] &
          [mane,X7] &
          [golden,X7] &
          [have,X4,X7]
          anchor
   ]
   anchor
       X1 to t1
       X2 to s1
       X3 to john1
&
   loc(t1,s1) overlap loc(now,here)
```

89 ms.

In this example it is probably clear that **it** finds its antecedent in the embedded context of the preceding sentence. Another point regarding the choice of interpretation is illustrated by the Ortcutt example given later; here a de dicto interpretation of at least one of the beliefs is required in order to avoid contradiction on the level of the believer. Taken together, we have the following three important principles:

1. The antecedent of a pronoun does not have to refer to anything in the real world, if it occurs in an opaque context.

2. Preservation of consistency within belief worlds (*cognitive coherence*) is to be preferred.

3. The preferred interpretation may depend on the context, i.e. the (desirable) principle of compositionality has to be given up on.

3.1 Beliefs

This section considers some aspects of the treatment of *beliefs*. In the first subsection an example of de re and de dicto interpretations is given, followed by a discussion of some axioms about knowledge and belief. In the following subsection *conflicting beliefs* are discussed, illustrated by the Ralph and Ortcutt case.

3.1.1 Inference Based on Beliefs

Let us first consider the following single sentence discourse: **"John believes that a president is powerful"**. The value-loaded interpretation gives us a specific individual about whom John believes that he is powerful, and the value-free interpretation describes John's belief that being a president is a powerful role.[5] Let us first show the **CDS** structure for this sentence given a referential de re interpretation, i.e. a value-loaded interpretation:

```
CDS structure:

    loc(X1,X2)
        [John,X3] &
        [president,X4] &
        [believe,X3,[powerful,X4]]
        anchor
            X1 to t1
            X2 to s1
            X3 to john1
            X4 to president1
&
    loc(t1,s1) overlap loc(now,here)

& ms.
```

The attributive de dicto interpretation, i.e. the value-free interpretation, looks like:

---

[5] Being a degree adjective, "powerful" should really be treated as a focalizer [McCord 1987]. For the sake of simplicity we shall disregard that in this example.

```
CDS structure:

   loc(X1,X2)
      [John,X3] &
      [believe,X3,[powerful,X4 in
         loc(X5,X6)
            [president,X4]
            anchor
                  ]
      ]
      anchor
         X1 to t1
         X2 to s1
         X3 to john1
   &
   loc(t1,s1) overlap loc(now,here)

9 ms.
```

In the referential interpretation any other reference to the president may be substituted for **X4** in the embedded statement. In the attributive de dicto interpretation nothing can be substituted for **X4**, since **X4** appears in an unrealized situation type describing John's way of referring to the president. But if John is told that **X** is a president and asked[6] if he thinks that **X** is powerful, John should conclude that the answer is yes. Or, in other words, if John believes that **X** realizes the role **X4 in Sit**, he should conclude that what is predicated of **X4** is also predicated of **X**. For an entity to realize a role **R** it must *unify* with the role, in our terminology be *anchored* to **R in Sit**, and **Sit** must be true.[7]

Generally, if **A** believes that a predication **Pred** holds of a role **X in Sit**, and **A** believes that an entity **Y** is described by that role, or, in other words, believes that **X in Sit** applies to **Y**, then **A** also believes that **Pred** is true of **Y**.

This can be stated by the following rule:

```
[Loc,[believe,A,[Pred,(X in Sit)]],Anchor] &
[Loc,[believe,A,(X in Sit) applies_to Y)],Anchor1] =>
[Loc,[believe,A,[Pred,Y]],Anchor1].
```

---

[6] We probably cannot generally assume that John will draw this conclusion unless prompted by something, but that is a matter we shall disregard here.

[7] This principle corresponds to the principle of *weak substitution* of Situation Semantics: *If t1 and t2 are each given the same interpretation (referential, outer attributive, or inner attributive throughout), then: If R believes that φ(t1) and R believes that t1 is t2, then R believes that φ(t2).* [Barwise and Perry 1983, p. 203-204 and p. 195]. According to them, weak substitution applies not only to beliefs, but also to knowledge, assertions, and primary and secondary see-reports.

The second condition on the antecedent side:

```
[[Loc,[believe,A,((X in Sit) applies_to Y)],Anchor]
```

is defined by:

```
[Loc,[believe,A,((X in Sit) applies_to Y)],Anchor] ←
        anchoring(anchor([X to Y]) &
        [Loc,[believe,A,Sit],Anchor] &
        anchoring(Anchor).
```

This rule says that somebody A believes that the role X as described in Sit applies to Y if A believes Sit with X and Y unified by anchoring. Of course, this rule can be generalized to cover all the cases where weak substitution applies.

Other axioms about knowledge and belief are common, but dangerous, if they involve self-reference and if they are expressed in a type-free language that makes no distinction between object and meta language. [Montague 1960] used the so-called *Hangman paradox*[8] to show that the following axioms about knowledge taken together will produce inconsistency (if just one is removed the inconsistencies do not arise):

M1. If you know ϕ, then ϕ.

M2. You know that if you know ϕ, then ϕ.

M3. If ϕ implies ψ and you know that ϕ, then you also know that ψ.

[Thomason 1980] shows that similar principles for beliefs also lead to inconsistency:

T1. If you believe that ϕ implies ψ and you believe ϕ, then you believe that ψ.

T2. You believe any theorem.

T3. If you believe that ϕ then you believe that you believe that ϕ.

T4. You believe that your beliefs are true.

However, some of these axioms require a degree of self-awareness that it does not seem reasonable to assume generally. E.g. the antecedents of M3, of T1 and of T3 can perfectly well be true without the agent actually drawing the inference. A Socratic way of teaching may be

---

[8]   A judge decrees on Sunday that a prisoner shall be hanged at noon of the following Monday, Tuesday, or Wednesday, that he shall not be hanged more than once, and that he shall not know until the morning of the hanging the day on which it will occur. It appears both that the decree can be fulfilled, and that it cannot.

A simpler version of this paradox is the *Crocodile's Dilemma*: A crocodile has stolen a child. The crocodile promises the child's father to return the child, provided that the father correctly guesses whether the crocodile will return the child or not. What should the crocodile do, if the father guesses that the crocodile will not return the child? [Kleene 1950].

seen as an effort to make people draw the inferences. Also T2 seems to assume too much. The reasoning behind T2 is probably that you know all theorems (the converse of M1, which is also commonly accepted) and you will probably believe anything you know and so you will believe all theorems. But it seems unreasonable to say that you know everything that is provable, or even worse, that you are aware of all the consequences of your knowledge and beliefs, as implied by M3 and T1.

For those reasons we shall not consider implicit inferences like M3 etc., but demand that the beliefs etc. be explicitly stated. In the rule about beliefs involving realizations of roles, this principle shows as the explicit belief held by A that the role X in Sit applies to Y.

We shall, however, accept the validity of T4, since it seems rather fundamental: you would certainly believe that your beliefs are true, otherwise you would not have those beliefs. This principle may in our formalism be stated as:

```
[Loc,[believe,A,([Loc,[believe,A,Sit,Anchor] => Sit)]),Anchor1]].
```

Note that the locations of the outer and embedded beliefs are the same; it seems reasonable to say that A believes that what he believes here and now is true, as opposed to what he today may believe about his beliefs yesterday.

3.1.2 Conflicting Beliefs

In this section we shall describe how conflicting beliefs may be handled in our system. It is very possible for a person to have conflicting beliefs about the same object. The believer does not recognize the identity of what he considers two different objects, and is thus able to ascribe conflicting attributes to them. Situation Semantics has coined the terms *cognitively coherent*, but *externally incoherent*, beliefs to this case.

Two famous examples of this concern Kripke's Pierre and Quine's Ralph and Ortcutt, respectively. Pierre grew up in France, learning that "Londres est jolie". Later, Pierre moves to London, a city he finds most disgusting, while not realizing the identity of Londres and London [Kripke 1979]. The other example is about Ralph who meets a suspicious looking man on the beach and believes him to be a spy, while not realizing that he is in fact identical to Ortcutt, whom he believes to be a respectable citizen [Quine 1956].

As will be seen from these examples, the beliefs are about the same thing in a strictly *external* sense, and so there is no conflict in the believer's mind, only on the level of the "real" world.

Let us consider the following version of the Ortcutt example:

```
Ortcutt is a spy. Ralph sees an old man. Ralph believes that the old man is a spy.
Ralph believes that Ortcutt is not a spy. The old man is Ortcutt.
```

The two first sentences, not involving beliefs, are fairly straightforward. Each of these two sentences introduces a situation in the usual way. For sentences three and four the question of what interpretation to give the beliefs comes up. Clearly, the beliefs cannot be given a *de re* interpretation, because that would introduce a contradiction. A *referential de dicto* interpretation, on the other hand, seems useful: Ralph believes of the person he thinks of as an old man that he is a spy.[9] Of the person Ralph thinks is Ortcutt, he believes that he is definitely not a spy.[10] It would probably be reasonable to give the beliefs of both these sentences a referential de dicto interpretation. For the sake of example, however, we shall give the belief about Ortcutt a referential de re interpretation.

The connection between the old man and Ortcutt is introduced by the final sentence. In the semantic representation this identity is given by an identity statement, which through the anchorings tie the old man and Ortcutt together. The identity relation can be used by the following rule of inference: If P can be proved of m and m is identical to n, then P can be proved of n. This introduces referential transparency on the level of the "real" world for identical referents referred to by non-identical reference markers. In this example, **Ortcutt** and **the old man** give rise to non-identical reference markers (**ortcutt1** and **man1**, respectively) and the identity of the two persons is introduced by anchoring the variables of the identity relation to **ortcutt1** and **man1**, respectively.

All in all, the **CDS** structure produced for this discourse is as follows:

```
CDS structure:

    loc(X1,X2)                      /* Ortcutt is a spy */
       [Ortcutt,X3] &
       [spy,X3]
       anchor
          X1 to t1
          X2 to s1
          X3 to ortcutt1
&
    loc(t1,s1) overlap loc(now,here)
&
    loc(X4,X5)                      /* Ralph sees an old man */
       [Ralph,X6] &
       [man,X7] &
       [old,X7] &
       [see,X6,X7]
       anchor
          X4 to t2
```

---

[9]   We shall assume that a direct perception indicated by e.g. **see** gives rise to the way the de dicto belief is formulated in the believer's mind, if nothing else is stated.

[10] Note here that our decision about how to treat names makes a de dicto interpretation of names possible.

```
            X5 to s2
            X6 to ralph1
            X7 to man1
&
    loc(t2,s2) overlap loc(t1,s1)
&
    loc(X8,X9)                          /* Ralph believes that  */
        [believe,X10,[spy,X11 in        /* the old man is a spy */
            loc(X12,X13)
                [man,X11] &
                [old,X11]
                anchor
                    X11 to man1]
        ]
        anchor
            X8 to t3
            X9 to s3
            X10 to ralph1
&
    loc(t3,s3) overlap loc(t2,s2)
&
    loc(X14,X15)                        /* Ralph believes that   */
        [believe,X16,[spy,X17] => false] /* Ortcutt is not a spy  */
        anchor
            X14 to t4
            X15 to s4
            X17 to ortcutt1
            X16 to ralph1
&
    loc(t4,s4) overlap loc(t3,s3)
&
    loc(X18,X19)                        /* The old man is Ortcutt */
        [identical,X20,X21]
        anchor
            X18 to t5
            X19 to s5
            X21 to ortcutt1
            X20 to man1
&
    loc(t5,s5) overlap loc(t4,s4)

227 ms.
```

Among the various questions it is possible to ask regarding this discourse we shall consider the following:

1. **Does Ralph believe that the old man is a spy?**

2. **Does Ralph believe that Ortcutt is a spy?**

3. **Is the old man a spy?**

The first two questions show something about what can and cannot be inferred about Ralph's beliefs, and for each of them we want to consider the different possibilities of interpretations of the embedded clause. The last question shows something about "reality" as the reader of the discourse, in this case the computer system, understands it.

Let us first consider the question: **"Does Ralph believe that the old man is a spy?"** The information in the discourse about Ralph's beliefs about the old man was given a referential de dicto interpretation, so it should come as no surprise that if we give the question the same interpretation we get the following answer:[11]

```
The following answer was found:

    loc(X1,X2)
       [believe,X3,[spy,X4 in
           loc(X5,X6)
              [man,X4] &
              [old,X4]
              anchor
                  X4 to man1]
       ]
       anchor
           X1 to t5
           X2 to s5
           X3 to ralph1

370 ms.
```

Let us note in passing that the anchoring of the location coordinates reflects that the reference point for a question is taken to be at the end of the current discourse and that a predication P holds at loc(T,S) if P can be directly inferred from the situation taking place at loc(T,S), or if there is a situation at a location Loc1 before (or overlapping) loc(T,S) where P holds, and nothing has happened in between Loc1 and loc(T,S) that contradicts this.

But what happens if we give the question a referential de re interpretation? Our claim here is that if somebody has a referential de dicto belief, then that person also has that belief on a de re interpretation, i.e. we may leave out the description of how the agent refers to that entity, as long as the reference is to the same object and as long as we do not assume any unstated, new ways of referring. So, the following axiom should be part of our semantics:

```
[believe,X,[P,Y]] ←
        [believe,X,[P,Y in [Loc,Sit,Anchor]]] &
        anchoring(Anchors).
```

---

[11] The LODUS system does not *generate* text, so positive answers are given in the form of CDS structures.

The reasoning behind this is as follows: A referential de dicto interpretation of a belief describes a belief about a particular object, and also how the believer refers to that object. So the believer actually has that belief about that object, no matter how he refers to it. However, the referential de dicto interpretation does *not* automatically licence other referential de dicto interpretations of the same referent, but referred to by a different description.

On the referential de re interpretation of the question the answer is as follows:

```
The following answer was found:

  loc(X1,X2)
    [believe,X3,[spy,X4]]
    anchor
        X1 to t5
        X2 to s5
        X3 to ralph1
        X4 to man1

388 ms.
```

The third possibility, to give the question an attributive de dicto interpretation, does not really make sense, since the definite description in this case would prefer to have a referent.[12]

Let us next consider the second question: **"Does Ralph believe that Ortcutt is a spy?"** Ralph's belief about Ortcutt was given a referential de re interpretation. On the basis of this, it seems reasonable that also giving this question a referential de re interpretation should lead to the answer "**no**":

```
The answer is no.

4932 ms.
```

If, on the other hand, the question is given a de dicto interpretation, referential or attributive, we can no longer assume this answer. Since the discourse does not give us any information about the way Ralph refers to Ortcutt, we cannot conclude anything about Ralph's beliefs about Ortcutt on a de dicto interpretation:

```
No answer could be found.

6293 ms.
```

---

[12] A case in which it probably *would* make sense to ascribe an attributive de dicto interpretation concerns the case of non-existing entities as in: **Does John believe that the Devil is evil?**

This illustrates that the previously stated axiom about the relation between a de dicto and a de re interpretation only holds in one direction.

Let us finally consider the last question: "**Is the old man a spy?**" This question refers to the actual facts about the world, as described by the discourse, and not to what is said is believed about the world. No matter which interpretation we give the definite noun phrase **the old man**, we want to be able to infer that the old man is a spy, on the basis of the information that Ortcutt and the old man are identical and that Ortcutt is a spy:

```
yes, because:
      [identical,man1,ortcutt1] at loc(t5,s5)
   &
      [spy,ortcutt1] at loc(t5,s5)

The following answer was found:

   loc(X1,X2)
      [spy,X3]
      anchor
         X1 to t5
         X2 to s5
         X3 to man1


   1425 ms.
```

The reason the interpretation we choose for the definite noun phrase does not matter is that the definite noun phrase does not appear in an intensional context.

4 Other Approaches

At this point it might be good to compare the **CDS** approach to the propositional attitudes with that of others'. The following three approaches are all relevant: *Intensional Logic*, *Situation Semantics*, and Landman's *Pegs and Alecs*. The problems generally involved in using Intensional Logic are stated in other places (e.g. [Schubert and Pelletier 1982], [Bernth 1988]) and we shall have nothing to add here. A comparison with Situation Semantics, on the other hand, seems useful since **CDS** does draw inspiration from this theory. Also the *Pegs and Alecs* theory is interesting because it deals with identity of referents based on information states, and not on the actual reference in the real world.

4.1 Situation Semantics

The Situation Semantics treatment of the attitudes has changed to some extent over the years. The strict Realism of [Barwise and Perry 1983] has softened a little as described in e.g. [Barwise 1988]. Still, Realism is at the heart of Situation Semantics, and therefore we shall mainly use the [Barwise and Perry 1983] version as the basis of our comparison.

Maybe the biggest difference between the early versions of Situation Semantics and **CDS** regarding the attitudes is the global aim. As one of its most important goals, Situation Semantics has the development of the underlying philosophy, particularly the introduction of a naive (in Davidson's sense; see below) and realistic treatment. **CDS** is more concerned with the *computational* aspect, even though we do consider philosophical issues as far as they are helpful in creating a computational semantics.

On a technical level, however, there are both similarities and differences. The similarities consist mainly in the introduction of some kind of complex variable consisting of a simple variable described by a situation type. In Situation Semantics the pair consisting of a variable and a describing situation type is called a *concept*. However, while the **CDS** role is used directly in the **CDS** structure for representing the way the agent refers, the Situation Semantics concept seems to be used only indirectly for handling *recognition* of entities, not in the actual representation of a discourse. A concept **o** is *applied* to an individual **b** if the agent has the belief that **o** and **b** are the same. The concept is the closest Situation Semantics comes to our way of handling the attitude verbs, but since [Barwise and Perry 1983] is very vague on this point, we shall have nothing more to say about this.[13]

The way of handling the attitude verbs in Situation Semantics reflects Barwise and Perry's desire for a naive treatment. According to [Davidson 1968] the meaning of a sentence should not be different just because the sentence happens to be embedded within the scope of an attitude verb. Embedded and non-embedded sentences should be treated the same way. In our view, this is questionable because it seems desirable to make it clear that we are concerned with the way the agent refers to something.

In CDS this is accomplished by the use of complex variables (roles) giving a description of the way of referring. Situation Semantics, on the other hand, just embeds a whole situation; i.e. the general format for a situation E involving a belief is:

$$E = \{<Loc,believe,A,E'(X);1>,<of,X,f(X);1>\}$$

where

---

[13]   If "concept" really means what is hinted at in [Barwise and Perry 1983], then we have something very far from the Realism that Situation Semantics wants so much.

$$E' = \{<Loc,p,X;1>\}$$

where E' is the believed situation, predicating something (here **p**) of the indeterminate X. The anchoring of X is given in the second state of affairs of E, where **f** is the anchoring function. In our view, it is much more natural to use a role. This difference in view about what is more natural may very well reflect a difference in the underlying philosophy. From a realist point of view, it does make some kind of sense to embed a whole situation instead of a complex variable. Another point is that, while Barwise and Perry refuse to speak in terms of de dicto and de re interpretations, this distinction is central to our treatment. The use of roles goes well with the idea of *describing* how the agent refers to something.

The above general format for a situation involving a belief or other attitudes, shows another difference between Situation Semantics and **CDS**, viz. the way the variable(s) appearing in the embedded statement can be anchored. As far as it is possible to gather from [Barwise and Perry 1983], there are the following two possibilities in Situation Semantics:

1. Not to anchor at all, in which case there is no referent in the real world (cf. our unicorn example given earlier). This seems very reasonable and corresponds to the **CDS** way of doing it.

2. To anchor it outside the describing situation. In this case the reference is to a specific thing existing in the real world.

The Situation Semantics way of handling the second case, i.e. the case where the reference is to something in the real world, apparently does not allow for any kind of distinction in whether the embedded statement is to be taken transparently or opaquely. The embedded situation is probably meant to describe how the agent refers to the object(s) in question (the opaque interpretation), and this makes it rather unclear how to represent the transparent interpretation. But, as mentioned above, this distinction between de dicto and de re interpretations is not really considered useful by Barwise and Perry, so maybe it is not quite fair to criticise them for apparently not being able to handle it.

Our treatment of the second case listed above differs slightly in that we chose to do the anchoring *within* the role. That is, we chose to do the value-loading on the level of the describing role. It is not clear how important it is to do the value-loading within the role; a distinction between value-loading within or outside the describing situation could probably be used to reflect something about the agent's awareness of the anchoring, but this is something we shall not go into here.

A third possibility provided by **CDS** is not to use a role at all, but directly insert a condition like [woman,X4] & [believe,X3,[smile,X4]] and then do the anchoring outside the describing situation. In this case we have the information that the believer has this belief about a specific thing in the real world, but no information about how the believer refers to the object in question. This is the transparent case.

In later versions of Situation Semantics, e.g. [Barwise 1988], the need for taking the agent's perspective into consideration is recognized (considering the emphasis on *situatedness*, it is somewhat surprising that this did not get a stronger emphasis earlier!). According to this paper,

> *Facts are relative to the focus situation s of concern to a situated, cognitive agent. Basic propositions classify the agent's classification of the focus situation. Hence a basic proposition p has two components: the focus situation s it concerns and the state of affairs σ it uses to classify s. Such a proposition is true if σ correctly classifies s; otherwise it is false.*

and

> *Focus situations can be determined by the cognitive activity of a situated agent, hence are relative to that agent's perspective.*

In the earlier versions of Situation Semantics, the dependence of the content on the perspective of the speaker was handled by trying to locate the relevant perspectival parameter in the descriptive content. Realizing the inadequateness of this, however, [Barwise 1988] suggests trying to locate the perspectival parameter in the situation the statement concerns; i.e. to allow the agent's perspective to help determine what the focus situation is.

What is here called *perspective* is clearly related to what is called the agent's *belief world* or *assumptions* in this paper, and thus is a move away from Realism towards what we call de dicto interpretations. Realism is preserved in the sense that situations are considered to be parts of the world, but they can be characterized in different ways according to the perspective of the agent.

The details of this approach still need to be spelled out, but as far as can be seen from the current descriptions, the treatment of propositional attitudes by Situation Semantics is getting closer to that of **CDS**.

4.2 Pegs and Alecs

[Landman 1986] introduces *pegs*, which are partial objects, on which information states can hang properties. Pegs may be indiscernable at some point, but discernable at another, depending on the information present. Two pegs are strongly indiscernable on the basis of the present information state if they have the same properties in it and if the properties incompatible with one peg are the same as those incompatible with the other peg. This definition clearly reflects the fact that it is possible to go in both directions when information is added: from discernablility to indiscernability as in the Ortcutt case if Ralph learns that the old man really is Ortcutt, or from indiscernability to discernability as in the case of two identical twins if you learn that one has a scar on his hand.

Identity of pegs is thus heavily based on the ability to tell them apart, and this in turn is based on *information*, and not on the actual reference in the real world, or in possible worlds. Here we shall not go into the concept of *alecs*, apart from mentioning that they are a special kind of pegs which, relative to some information state, can play the role of pegs with certain properties.

The idea of being able or unable to discern certain objects on the basis of information is clearly related to our concept of roles. In our treatment of opaque contexts, a role is used to describe the way the agent of the attitude refers to the entity in question, or, in other words, the role describes (parts of) the information that the agent has about that entity. A very interesting issue we do not intend to address here is how to keep the mental states of an agent consistent. As the agent's state of information changes, new information may require deletion of old information in order to maintain cognitive coherence. One way of dealing with *revision of beliefs* is described in [Foo and Rao 1988].

Acknowledgements

I would like to thank Michael McCord for letting me use his parser and for commenting on earlier versions of this work, Dan Sahlin and Seif Haridi for letting me use and modify their intuitionistic theorem prover, and Manny Rayner for useful discussions on my approach to opaque contexts.

References

Barwise, J. [1988] "Situations, Facts and True Propositions", in Barwise, J. [1989] *The Situation in Logic*, CSLI Lecture Notes No. 17, Chicago University Press.

Barwise, J. and J. Perry [1983] *Situations and Attitudes*, MIT Press.

Bernth, A. [1988] *Computational Discourse Semantics*, Ph.D. Thesis, University of Copenhagen.

Bernth, A. [1989] "Discourse Understanding In Logic", *Proc. North American Conference on Logic Programming*, MIT Press.

Davidson, D. [1968] "On Saying That", *Synthese* vol. 19 (1968-69).

Donnellan, K. [1966] "Reference and Definite Descriptions", *Philosophical Review*, 75 (1966).

Evans, G. [1977] "Pronouns, Quantifiers and Relative Clauses (I)", *Canadian Journal of Philosophy* 7, pp.467-536.

Foo, N.Y. and A.S. Rao [1988] "Belief Revision in a Microworld", Technical Report 325, Basser Department of Computer Science, University of Sydney.

Haridi, S. and D. Sahlin [1988] "Logical Extensions of Logic Programming Based on Intuitionistic Logic", forthcoming SICS Research Report, Swedish Institute of Computer Science.

Kamp, H. [1981] "A Theory of Truth and Semantic Representation", in Groenendijk et al. (eds) *Formal Methods In the Study of Language*, Mathematical Centre Tract, Amsterdam.

Kleene, S.C. [1950] *Introduction To Metamathematics*, van Nostrand.

Kripke, S. [1979] "A Puzzle about Belief", in Margalit (ed): *Meaning and Use*, Reidel.

Landman, F. [1986] "Pegs and Alecs", in Landman, F. [1986] *Towards a Theory of Information. The Status of Partial Objects in Semantics*, Foris.

McCord, M.C. [1981] "Focalizers, the Scoping Problem, and Semantic Interpretation Rules in Logic Grammars", in van Caneghem, M. and D. H. D. Warren (eds) [1986] *Logic Programming and its Applications*, Ablex.

McCord, M.C. [1987] "Natural Language Processing in Prolog", in Walker, A. (ed), M.C. McCord, J. Sowa and W. Wilson [1987] *Knowledge Systems and Prolog: A Logical Approach to Expert Systems and Natural Language Processing*, Addison-Wesley.

McCord, M.C. [1989] "A New Version of Slot Grammar", Research Report RC 14506, IBM Research Division, Yorktown Heights, NY 10598.

McCord, M.C. [1990] "SLOT GRAMMAR: A System for Simpler Construction of Practical Natural Language Grammars", Research Report RC 15582, IBM Research Division, Yorktown Heights, NY 10598.

Montague, R. [1960] "A Paradox Regained", in Thomason, R. (ed) [1974] *Formal Philosophy, Selected Papers of Richard Montague*, Yale University Press.

Quine, W.V. [1956] "Quantifiers and Propositional Attitudes", *Journal of Philosophy* vol. 53, pp. 177-187.

Richards, B. [1981] "Pronouns, Reference and Semantic Laziness", in Heny, F. (ed) [1981] *Ambiguities in Intensional Contexts*, Reidel.

Schubert, L.K. and F.J. Pelletier [1982] "From English to Logic: Context-Free Computation of 'Conventional' Logical Translation", *American Journal of Computational Linguistics* vol. 8 no. 1 1982.

Knowledge Processing in the LILOG Project
-
From the first to the second Prototype

Toni Bollinger

Karl-Hans Bläsius

Uli Hedtstück

IBM Deutschland GmbH
Wissenschaftliches Zentrum
Institut für Wissensbasierte Systeme
Schloßstr. 70
D - 7000 Stuttgart
Fed. Rep. of Germany

FH Rheinland-Pfalz
Abteilung Trier
Fachbereich 3
Schneidershof
D - 5500 Trier
Fed. Rep. of Germany

FH Konstanz
FB Wirtschaftsinformatik
Brauneggerstr. 55
Postfach 7729
D - 7750 Konstanz
Fed. Rep. of Germany

Abstract

This paper gives an overview of the knowledge processing in the LILOG project. We describe the knowledge processing component as it has been realized within the first prototype and give some experimental results. The experiences gained with the first prototype motivated us to improve the knowledge processing for the second prototype. These modifications are presented in the second part of this paper.

1. Introduction

LILOG (LInguistic and LOGical methods) is a project which is currently undertaken by IBM Germany at Stuttgart together with some university partners. The aim of LILOG is the investigation of linguistic and logical tools and methods for the computational understanding of texts. Understanding of text means the construction of a semantic representation of a piece of text, that is a (partial) model of the situation described in the text.

In 1987 a prototype was implemented able to understand a German text (taken from a tour guide) describing a hiking tour in the Alsace and to answer questions about the domain and the text. The knowledge representation language L_{LILOG}, which is based on predicate logic, has been developed to represent the different kinds of knowledge necessary for the process of text understanding as well as to represent the semantic description of the input text. A knowledge processing component has been implemented as interpreter for L$_{LILOG}$ able to solve problems specified in L$_{LILOG}$ and to extend the knowledge base deriving new facts.

On the basis of the experience made with the first prototype the representation language L$_{LILOG}$ has been extended to L$_{LILOG}$-II and the demands on the knowledge processing component have been enlarged too. A second prototype is currently being developed taking into account the new requirements and the experience made with the first prototype.

This paper gives an overview of knowledge processing in the LILOG project. We describe the knowledge processing component (also called inference machine) as it has been realized within the first prototype and give some experimental results. Then we explain the knowledge processing component for the second prototype which is currently being developed.

We expect the reader to be familiar with fundamentals of artificial intelligence, especially knowledge representation formalisms, as well as predicate logic and the resolution calculus.

2. Knowledge Processing in the first LILOG-Prototype

The first LILOG-prototype is a text understanding system which has additional question answering capabilities for determining the degree of understanding of a text. The input to the system can be a text or a question. The linguistic processing performs parsing and generates a semantic representation of the input sentence or question.

For every new sentence the corresponding representation is integrated into the knowledge base by the knowledge processing component which after storing the facts representing the meaning of the sentence, essentially performs consistency tests and derives further facts by forward inferences.

Questions are answered by trying to deduce their representation from the existing knowledge. The results are then passed to the language generation component which formulates the answer.

In this section we will at first briefly describe the knowledge representation language L$_{LILOG}$ which is used for formulating the rules and facts of the background knowledge. Then we give a detailed description of how a sentence or a question is processed and will present the most important types of inferences on L$_{LILOG}$ expressions. The section ends with an account of our experiences with the first prototype.

2.1 The Knowledge Representation Language L_{LILOG}

From a logical point of view L_{LILOG} (Beierle et al. 88) can essentially be identified with order-sorted Horn logic. Order-sortedness means that with every term a sort is associated that can be chosen among a predefined set of sorts. The set of sorts is partially ordered and is a lower semilattice in order to guarantee the existence of greatest lower bounds. Due to the restriction to Horn logic a knowledge base in our system is composed of two types of formulas, called **knowledge elements** in the L_{LILOG} terminology. These are:
- facts, being single literals, and
- rules, which are (special) implications between a conjunction of literals and a single literal.

Since knowledge elements consist of literals, negated atoms are allowed to occur there too. But on the basic inference level (cf. the inference rules mentioned in section 2.3), the negation is not exploited logically, i.e., negated predicates are considered there as new predicates with no relation to their positive counterparts.

Another peculiarity of L_{LILOG}, is that arguments of predicates are not identified by their position, but by roles. While this is only "syntactic sugar" from the logical point of view, roles may have a certain importance for the linguistic processing.

There is also a special type of constants used for entities determined during the analysis of a text. They are called **reference objects**. For example the knowledge element
    see(agent = r1:HUMAN_BEING, object = r2:HOUSE)
represents the fact that the human being r1 sees the house r2. "Agent" and "object" are the roles of the predicate "see". r1 and r2 are reference objects. HUMAN_BEING and HOUSE are the sorts declared for r1 and r2.

A particular feature of L_{LILOG} is that sorts can also be used as special unary predicates. While sorts in a sorted logic are usually associated with terms in a static way, we have introduced sorts as predicates in order to be able to reason explicitly about the sort membership of a term. More about this will be said in the subsequent sections. An atomic formula whose predicate is a sort is called a sort literal or sort formula.

Entrypoints are used to specify control information. If a premise of an implication is marked by an entrypoint, it can be used for forward inferences, if the literal in the conclusion is marked we have a backward chaining rule. Rules have to contain at least one entrypoint for being used during an inference process. Implications without entrypoints are completely useless.

The following rule is an example of a backward chaining rule, as its conclusion is marked by an entrypoint EP. Transscribed into natural language it means that if a human being X drives a vehicle Y then he moves along with it.

```
ke1: forall X: HUMAN_BEING, Y: VEHICLE
      ( drive( agent = X, medium = Y )
        impl
        EP move( agent = X, medium = Y ) ).
```

The rule ke2 can be used for forward inferences. It states that everything being narrow is not broad:

```
ke2: forall X:ALL
      ( EP narrow( subject = X ) )
        impl
        not( broad( subject = X ) ) ).
```

Note here that the negated predicate in the conclusion has to be considered as a "complex" predicate.

Knowledge elements taken by default are marked by a special **default marker**. One can represent for example the assertion that castles typically have a tower by the following L$_{LILOG}$ formula:

```
forall C:CASTLE exists T:TOWER
      ( have(agent = C, object = T) DEFAULT ).
```

Knowledge bases can be structured into **knowledge packets** (Wachsmuth 1987). The set of knowledge packets is partially ordered and allows a partioning of a knowledge base into visible, reachable and nonreachable parts. Further, in order to distinguish between the system's a priori world and domain dependent knowledge and the knowledge acquired from a text, knowledge bases are supplied with the types "background" or "text".

2.2. Inferencing Tasks

For sentences and for questions the output of the linguistic processing is a discourse representation structure (DRS, Kamp 1981). A DRS can be interpreted as a quantified logical formula. This formula is first transformed into a conjunctive normalform and then skolemized before the proper processing begins.

If there are multiple readings, a list of DRSs is created that is treated differently if it represents the meaning of a sentence or a question.

2.2.1 Inferencing Tasks for Text Understanding

In the text understanding mode the first reading is taken arbitrarily as the appropriate one. Normally a DRS represents a conjunction of facts. Conjuncts can also be implications, but these are only stored in the text knowledge base without performing consistency checks and forward inferences.

For conjunctions of facts the following inferencing tasks are performed:
- consistency checks,
- forward inferences and
- storage of sort literals.

2.2.1.1 Consistency Checks

For every new fact extracted from the text it is tested if the negation of this fact can be deduced from the rules and facts in the background and text knowledge base. If such an inconsistency has been detected the corresponding fact is marked as being contradictory and stored in the text knowledge base.

Every fact is stored, even the contradictory ones, as we want all the knowledge extracted from a text to be contained in the text knowledge base. If we find a contradiction, we contend ourselves to mark it, instead of resolving the conflict in a more sophisticated way.

Example:

In the presentation of the examples we state only the English translations of the German sentences and questions. The examples are also simplified to clarify the presentation.

Let us suppose that the following two facts are contained in the text knowledge base:
$$te1: \quad narrow(\text{ subject } = r_{gr}:\text{STREET}),$$
$$te2: \quad not(broad(\text{ subject } = r_{gr}:\text{STREET})) .$$
Te1 is one of the facts created for the sentence "The narrow Grand' Rue is situated in Saverne."; r_{gr} is the reference object for the street "Grand' Rue". Te2 can be inferred by forward inferences using the forward chaining rule ke2 in section 2.1.

If one enters the sentence: "The Grand' Rue is broad.", the fact
$$te3: \quad broad(\text{ subject } = r_{gr}:\text{STREET})$$
is created, which obvisiously contradicts te2. Therefore, it gets a contradiction label.

2.2.1.2 Forward Inferences

For every new fact without contradiction label all forward inferences within a certain limit are performed. This is done by unifying a new fact with a premise literal marked by an entrypoint and then trying to prove the remaining premises of the forward chaining rule. If this succeeds, the correspondingly instantiated conclusion is a valid fact. From these facts new facts are deduced by forward inferences too. This cycle is iterated several times. The maximal iteration depth is specified by a global parameter of the system.

Example:

Let us suppose that the background knowledge contains the following rule expressing the transitivity of the predicate loc_in (meaning "is located in"):

ke3: forall A1 : ALL, A2 : ALL , A3 : ALL
    ((EP loc_in(arg1 = A1, arg2 = A2)
       and
      loc_in(arg1 = A2, arg2 = A3))
    impl
    loc_in(arg1 = A1, arg2 = A3)) .

ke3 is a forward chaining rule as the first premise is marked by an entrypoint EP. Further, let the following text element be in the text knowledge base

te4: loc_in(arg1 = r_{gr}:STREET , arg2 = r_{sav}:TOWN)

(which means "The Grand' Rue is located in Saverne", where r_{sav} is the reference object for Saverne.)

The meaning of the sentence
    "The hiking tour begins in the Grand' Rue."
is represented by the conjunction of the facts te5 and te6:

te5: begin(subject = r_{ht} : HIKING_TOUR , aspect = r_{ev} : EVENT),
te6: loc_in(arg1 = r_{ev} : EVENT, arg2 = r_{gr}: STREET).

Te5 states that the event of the beginning of the hiking tour is r_{ev}, and the statement that this event is located in the Grand' Rue is expressed by te6. Now, a forward inference step can be performed using te6 and ke3, because te6 and the premise of ke3 marked by an entrypoint are unifiable. As the second premise can be proved by te4, the fact te7 is deduced:

te7: loc_in(arg1 = r_{ev}:EVENT, arg2 = r_{sav} : TOWN)

indicating that the beginning of the hiking tour is located in Saverne, too.

2.2.1.3 Storage of Sort Literals

For every reference object r with sort restriction S occuring in the respresentation of a sentence, the sort formula S(element = r:S) is created and stored in the text knowledge base. This is done for answering questions about the sort membership of reference objects (cf. section 2.3.4) and to have an explicit logical representation of this relation.

It may happen that due to new information the sort of a reference object is specialized during the text understanding process. This change of the sort is recognized when the corresponding sort literal is stored in the text knowledge base. The sort of the reference object then has to be updated in every formula it occurs in, as otherwise valid inferences could be excluded due to unsatisfied sort restrictions.

Example:

At the beginning the reference object r1 has the sort HUMAN_BEING. Hence the sort formula
      HUMAN_BEING(element = r1:HUMAN_BEING)
is stored.
Further in the text the system learns that r1 is hiking. By forward inferences it can be deduced that r1 belongs to the sort HIKER. After storing
      HIKER(element = r1:HIKER)
the sort of r1 is changed to HIKER in all formulas.

2.2.2 Inferencing Tasks for Answering Questions

Questions are used for checking the contents of the background and text knowledge base. Therefore they are another means of testing how a text has been understood by the system.

There are two kinds of questions that can be posed to the LILOG system:
- yes/no-questions: "Does the hiking tour begin in Saverne?",
                     "Is Haut-Barre a sight?"
- wh-questions: "Where does the hiking tour begin?",
                     "Which sights do exist?"

It may occur that a question has several readings. In that case one reading is treated after the other until an answer has been found.

The representation of a question is mostly a conjunction of facts which we will call the goal. Questions are tried to be answered by looking for a proof of the corresponding goal. Proofs using contradictory facts are not considered unless there are no other proofs. In such a case the information is given to the language generation that an answer could only be found on the basis of inconsistent knowledge.

Different actions are performed if a yes/no-question or wh-question is to be answered.

2.2.2.1 Yes/No-Questions

For yes/no-questions, if a proof of the goal has not been found, or if there have only been proofs using defaults, then an additional proof of the negated goal is tried. The answer is given according to the criterion that nondefault proofs win over default proofs, i.e., e.g., if there is a proof of the goal using defaults, but a proof of the negated goal without defaults, the answer is "no". In case that there are only default proofs of the goal and its negation, the answer is an "optimistic" "typically yes", even though answering "typically no" would be justified as well.

The following table summarizes the answers to yes/no-questions:

proof of NOT(G) \ proof of G	without defaults	with defaults	no proof
without defaults	yes •	no	no
with defaults	yes •	typically yes	typically no
no proof	yes •	typically yes	don't know

• NOT(G) not tested

Table1: Answers to yes/no-questions

Example:

The question "Is the Grand' Rue narrow?" is represented by the following goal:
  G1:   narrow(subject = r_{gr}:STREET).
It can immediately be proven by te1 (see section 2.2.1.1), leading to the answer "yes".

Before entering the sentence "The Grand' Rue is broad.", the answer to the question "Is the Grand' Rue broad?" with the goal
  G2:   broad(subject = r_{gr}:STREET) ,
would be "no" due to te2. After entering the contradictory sentence, G2 can only be proved by te3, which has a contradiction label. So the system replies that only an answer has been found that is based on contradictory information.

2.2.2.2 Wh-Questions

For answering wh-questions valid instantiations of a certain variable of the goal are searched. They are determined by trying to find (under certain restrictions) all proofs of the goal. Before passing these instantiations to the language generation, certain answers that are redundant or may seem superfluous to the user are dropped by the answer compression module. E.g, an answer found by a default proof and being represented by a skolem-term (standing for an existentially quantified variable) of sort S is considered as redundant, if there is a nondefault answer represented by a reference object belonging to a subsort of S. To give another example, considering the following answers to the question: "Which sights do exist?": "all sights", "the castle Haut-Barr" and "the Ruin Greifenstein", one realizes that the first one contains no relevant information. Therefore, it is eliminated by the answer compression module.

Example:

The question "Where does the hiking tour begin?" is represented by the following goal:
    begin(subject = r_{ht}:HIKING_TOUR, aspect = E:EVENT)
        and
    loc_in(arg1 = E:EVENT, arg2 = A:ALL).
The varibales E and A are implicitly existentially quantified. A is the variable whose instantiations are looked for.

There are two proofs of the goal, the first using the text elements te5 and te6:
    te5:   begin(subject = r_{ht}:HIKING_TOUR , aspect = r_{ev}:EVENT),
    te6:   loc_in(arg1 = r_{ev}:EVENT, arg2 = r_{gr}:STREET).
In this proof the instantiation of A is r_{gr} such that "in the Grand' Rue" is the answer. In the second proof te7 is used instead of te6:
    te7:   loc_in(arg1 = r_{ev}:EVENT, arg2 = r_{sav}:TOWN).
Since now the instantiation of A is r_{sav}, the second answer is "in Saverne". So the final answer of the system is "In the Grand' Rue and in Saverne".

2.3 The Basic Inference Rules

In the preceding section we have described the inferencing tasks performed in our system for natural language understanding. The realization of these inferencing tasks is based on two general inference strategies: During the input mode, when new text is analyzed, forward chaining inferences are used to infer additional knowledge from the new facts. For consistency checks and for answering questions backward chaining inferences are performed.

In this section we want to describe the basic inference rules which are implemented for performing the above inference strategies. The inference rules are variants of ordinary resolution and incorporate order-sorted unification. For the special treatment of sort formulas we use theory resolution (Stickel 1985).

2.3.1 Order-Sorted Unification

Dealing with terms which are sorted on the basis of a given lower semilattice of sorts, we have to extend the usual unification procedure by order-sorted unification. Order-sorted unification may be shortly described by the following conditions:

(a)   A term, which is not a variable, and a variable are unifiable if the sort of the term is a subsort of the sort of the variable.

(b)   Two variables unify, if the greatest lower bound of their sorts is not BOTTOM, the empty sort. In this case, the two variables are both replaced by a new variable which has this greatest lower bound as its sort.

For more details see (Walther 1987), (Schmidt-Schauß 1988), and for a general survey (Siekmann 1989).

2.3.2 Backward and Forward Chaining

Similar to PROLOG, a backward resolution step consists of unifying a goal, or one of the conjuncts of a conjunctive goal, with a fact or the conclusion of an implication. In the case of an implication the conclusion has to have an entrypoint. Unification is done by order-sorted unification. The result of such an inference step is the resolvent consisting of the remaining subgoals and premises of the implication, to which the unifying substitution has been applied. In order to prove the original goal backward resolution steps are iterated until the empty clause has been obtained.

For a forward resolution step, a unary fact has to be unified with a premise of an implication which is marked by an entrypoint. In order to draw the conclusion, all other premises have to be proven first by a backward chaining proof.

2.3.3 Theory Resolution for Sorts

We will now describe the special treatment of sort formulas by theory resolution (Stickel 1985). The aim is to incorporate specialized reasoning procedures into a resolution based theorem prover taking into account the information contained in the sort hierarchy given by the lower semilattice of sorts. With theory resolution it is possible to resolve pairs of literals whose predicates are not identical.

We implemented the following subsort-b-resolution rule. For brevity, we omit the roles. $S_i(t:S_j)$ denotes a sort formula where S_i is the sort used as unary predicate, and S_j is the sort of the term t. LG, RG and PR are conjunctions of literals and stand for left goal, right goal and premise respectively.

subsort-b-resolution:

LG and $S_3(t_2:S_4)$ and RG   (conjunctive goal)
PR impl EP $S_1(t_1:S_2)$ (knowledge element)
$S_1 \leq S_3$ (sort information contained in the sort hierarchy)
$\mu(t_1:S_2) = \mu(t_2:S_4)$ (order-sorted unification by the substitution μ)

$\mu($LG and PR and RG $)$ (subsort-b-resolvent)

A forward chaining rule "supersort-f-resolution" may be defined analoguously by taking into account the respective supersort relation.

Sort formulas containing terms which are not variables are treated by an additional backward chaining inference rule based on the second order axiom

$$\text{forall } S \quad (\text{ forall } X{:}S \quad S(X{:}S) \ )$$

which we call "sort axiom". We always apply it to a goal in addition to subsort-b-resolution. The axiom expresses the fact, that for any sort S and term t which is declared to be of sort S the sort formula S(t:S) holds. Taking into account the subsort relations given by the sort semilattice we get the following inference rule.

sort-axiom-resolution (for nonvariable terms):

$$\text{LG and } S_1(t{:}S_2) \text{ and RG}$$
$$S_2 \leq S_1$$

$$\text{LG and RG}$$

Goals which are sort formulas whose argument is a variable are treated differently. A variable in a goal is implicitly existentially quantified. This situation is given, for example, in the goal BUILDING(X:SIGHT) representing the question "Is there a sight which is a building?". Based on the convention that we only admit non-empty sorts the goal $S_1(X:S_2)$ is true whenever BOTTOM $<$ glb(S_1,S_2) (where glb is an abbreviation for "greatest lower bound").

For negated sort formulas we have implemented a corresponding sort-axiom-resolution rule.

2.3.4 On the Use of Sort Literals and the Sort Inference Rules

In section 2.2.1 we have already said that in the text understanding mode sort literals are created for every reference object and stored in the text knowledge base. We have further mentioned that the sort restriction of a reference object is updated if its sort has been specialized.

In the question answering mode sort literals and sort inference rules are used for answering two kinds of questions that concern the structure of the sort lattice and the sort membership relation of reference objects. These questions are
- the "exist-questions", like "Which sights do exist?", and
- the "is-questions", like "Is Haut-Barr a sight?" or "Is a castle a ruin?".

An exist-question, like "Which sights do exist?", is mapped to the following goal, where for brevity some of the arguments of the predicate "existing" are omitted and the variable X is implicitly existentially quantified:

G_{exist}: existing(X:SIGHT)

For proving G_{exist} the backward chaining rule ke0071 is used:

ke0071: forall A:ALL (ALL(X) impl EP existing(X)).

Together with the sort formula ke_{hb} created for the reference object r_{hb} of the castle Haut-Barr

ke_{hb}:   CASTLE(r_{hb}:CASTLE)

we get the following proof of the goal:

b-resolution of G_{exist} with ke0071: G'_{exist}:   ALL(X:SIGHT)
b-subsort-resolution of G'_{exist} with ke_{hb}
(as CASTLE is a subsort of SIGHT and ALL): G''_{exist}: |_| (empty clause)

The first answer is therefore "Haut-Barr".

G'_{exist} can also be proven by the sort-axiom-rule for variables as SIGHT is a subsort of ALL. That is how we obtain the answer "all sights", which is eliminated by the answer compression (cf. section 2.2.2.2).

If one looks closer at the rule ke0071 one could argue that the premise ALL(X:ALL) could be eliminated as it is always true and therefore redundant. From a logical point of view this is correct. But if one considers the consequences, one realizes that this would lead to pragmatically useless answers.

If we would eliminate the premise we would get the fact

ke0071':   forall X:ALL existing(X).

This fact suffices to prove the goal G_{exist}. But there would be no other proof so that the only answer would be "All sights". This answer represents logically the complete answer set, as for instance r_{hb} is also a member of the sort SIGHT. But it is obvious that for a human user this answer is not at all the one he wanted to have.

A reason for this phenomen is that in unification based calculi (e.g. the resolution calculus) only most general solutions are derivable, and it is not possible to derive arbitrary instances of a formula. However, in many cases the user does not expect to get only one most general, but rather some concrete individual solutions. With the method described above, our system also creates some instances as individual solutions by considering the reference objects, which have the sort compatible with the general solution. The general solution is eliminated by the answer compression module. Hence the system produces only some individual solutions as expected by the user.

These considerations show that reasoning in theorem proving may be different from reasoning in knowledge based or text understanding systems, where pragmatic considerations like the expectations of the user have also to be taken into account.

Is-questions are answered by using the sort-axiom-rule. The question "Is Haut-Barr a sight?" is represented by the following goal:

G_{hb}:   SIGHT(r_{hb}:CASTLE),

and the sort-axiom rule for nonvariable terms can be successfully applied as CASTLE is a subsort of SIGHT.

In an analogous way the existential reading of the question "Is a castle a ruin?", which means "Is there a castle which is a ruin?", is proven. It has the representation

G_{castle}:   RUIN(X:CASTLE)

and the sort-axiom rule for variables works here, as RUIN and CASTLE have the greates lower bound RUIN_OF_CASTLE.

2.4 Experiences with the first Prototype

Table 2 shows some figures about the background and the text knowledge after the 14 sentences of the hiking tour description have been processed:

257 predicates
285 sorts
307 rules
144 facts (background knowledge)
138 facts (text knowledge)
271 facts (deduced by forward inferences)
  4 reference objects (background knowledge)
154 reference objects (text knowlegde)

Table 2: Figures about the knowledge bases in the first prototype

Comparing the number of background and text knowledge elements (a total of 860) to the number of declared predicates (285) one realizes that on the average the number of knowledge elements per predicate is quite small. This contrasts to the situation with theorem proving applications where the ratio of axioms to predicates is greater but the number of axioms much smaller in general.

A consequence of this observation is that, although we have quite a high number of rules and facts, the search space for finding proofs remains managable. In fact, the proofs for answering questions were in general quite short (up to 5 rule applications) and this was partially due to this characteristic of the knowledge bases.

The number of sorts is quite high, as the meaning of nouns is respresented by sorts. Furthermore, for any two sorts having a greatest lower bound (nonempty intersection), the corresponding sort has to be predefined too and to be integrated into the sort structure. For more general applications with more extended dictionaries this could create a problem, as the sort lattice would grow to a hardly managable size. Therefore we decided to develop a sort description language for the second prototype which does not force us to introduce an atomic symbol for every new sort.

As expected, the use of sorts also had a positive effect on the length of proofs, as in unsorted logic a supersort-subsort-check would require several resolution steps.

Several drawbacks of L$_{\text{LILOG}}$ and the knowledge processing component have been detected. The most important is that Horn logic is not expressive enough for our application, especially for spatial and temporal reasoning But also the default reasoning capabilities were too poorly developed for handling the assertion and retraction of assumptions during the text understanding process.

3. Knowledge Processing in the second LILOG-Prototype

The experiences made with the first prototype led to new requirements for knowledge representation and knowledge processing. On the one side they concern the expressive power of the knowledge representation language, and on the other side they concern new problem solving tasks to be performed by the knowledge processing component. For example, the linguistic component in order to construct a semantic representation from a given input text should be supported by executing certain problem solving processes.

There were several requirements for an extension of the knowledge representation language L_{LILOG} like "full first order predicate logic", "equality reasoning", "a powerful sort description language", "default reasoning", "reason maintenance", "vagueness", "uncertainty", "arithmetic", "sets" and "generalized quantifiers". Most of them have been integrated into the new knowledge representation language L_{LILOG} II, which will be described briefly in section 3.1. The inference tasks to be solved by the inference machine are described in section 3.2. Section 3.3 gives an overview of the new knowledge processing component which is currently under development based on the knowledge representation language L_{LILOG} II.

3.1. The Knowledge Representation Language L_{LILOG} II

In this section we will give an informal introduction of the knowledge representation language L_{LILOG} II. A detailed description of this language can be found in (Pletat, v. Luck 1989). First we explain the sort description language $SORT_{L-LILOG}$ as one essential part of L_{LILOG} II, and then we explain the logical part and the extensions of L_{LILOG} II.

3.1.1. Sorts in L_{LILOG} II

3.1.1.1 The Sort Description Language $SORT_{L-LILOG}$

Our experiences with the first prototype have shown that we need more expressive power and more flexibility for the representation of sort information in the framework of a natural language understanding system.

For the description of sorts we developed the $SORT_{L-LILOG}$ formalism based on the work described in (Beierle et al. 1988) and on ideas coming from the KL-ONE family of knowledge representation formalisms (Brachman, Schmolze 1985).

One principle of SORT$_{L\text{-}LILOG}$ is that the complete sort hierarchy is not represented internally. Instead, sort descriptions are generated dynamically with the aid of given sort names and constructors.

The supply of symbols for SORT$_{L\text{-}LILOG}$ is given by a **sort signature** which contains those symbols over which **sort expressions** can be formed. The construction of sort expressions will start from basic sorts given in the set of **sort names** SN of the sort signature. **Features, roles** and **atoms** are further components of the sort signature. By means of constructors complex sort expressions can be formed.

More exactly, a sort signature S is a quadruple $S = < SN, F, R, A >$, where

- SN is a set of sort names, such that BOTTOM, TOP are in SN
- F is a set of features
- R is a set of roles
- A is a set of atoms.

The set SE(S) of sort expressions over S contains

- SN
- (se **and** se′)
- (se **or** se′)
- ¬se
- $\{a_1,...,a_n\}$
- **with** fp **in** se
- **with** r **in** se
- **some** r
- **agree** fp fp′
- **disagree** fp fp′

where se, se′ are in SE(S), r in R, a_i in A. fp, fp′ are feature paths, i.e., concatenations of features, such that set valued features only occur at the end.

3.1.1.2 Semantics for SORT$_{L\text{-}LILOG}$

Sort expressions are used in order to describe sorts which are interpreted by sets. Thus any sort signature $S = <SN,F,R,A>$ is given a set theoretic interpretation by interpreting the sort TOP by a non-empty universe, BOTTOM by the empty set, and the elements of SN by non-empty subsets of the universe. The elements of A are interpreted by single elements of the universe, expressions of the form $\{a_1,...,a_n\}$ denote finite subsets of the universe. Features are interpreted by functions, feature paths correspond to concatenations of functions, and roles are interpreted by relations. The operators and, or, ¬ are interpreted by intersection, union, and complement, respectively. Agreement and disagreement is interpreted by equality and inequality, respectively, for the sets which interpret the value sorts of the respective feature paths. For more details, in particular for the interpretation of the with and some operators, see (Pletat, v. Luck 1989).

3.1.1.3 Sort Hierarchies

Two sort expressions are equivalent if they are interpreted by the same set in any interpretation. The sort hierarchy for a given sort signature characterizes the structure on the set of corresponding equivalence classes of sort expressions. It is completely determined by the signature and the semantics for sort expressions inducing a Boolean lattice with respect to the ¬, and and or operations.

By means of the equivalence relation (in the following denoted by the symbol $\equiv$) it is possible to express constraints on the set of sort expressions.

For example, for a given sort signature S a partial ordering is defined on SE(S) by

$$se \leq se' \quad \text{iff} \quad (se \text{ and } \neg se') \equiv \text{BOTTOM}.$$

The relation denoted by $\leq$ is called the subsumption relation.

Another example is the constraint (se and se') $\equiv$ BOTTOM, which expresses, if two sort expressions denote disjoint sets.

3.1.2. Logic and Extensions

The knowledge representation language LLILOG II essentially is based on order-sorted predicate logic with equality (see e.g. (Cohn 1987), (Walther 1987), (Schmidt-Schauss 1988)). However, we not only have constant sorts, but also complex sort terms as described in the previous section. In the following we will explain some further extensions, most of which are motivated by linguistic requirements.

Just as in the first prototype we will use default markers, inconsistency markers, knowledge packets, reference objects, entrypoints and roles for identifying the arguments of predicates and functions (see section 2.1). However the treatment of these features will be improved in the future, exploring default reasoning with priority. The concept of entrypoints is currently being extended to a special control language allowing the formalization of knowledge (meta knowledge) about the knowledge (object knowledge) in the knowledge base. For the spatial reasoning, the control language will allow to switch to the depictional module which handles an analogue representation of spatial relationships (Khenkhar 1988). In general, the control language can be used to specify how the object knowledge is to be used by the inference machine. The concept for this control language has not yet been finished.

In the second prototype a restricted treatment of arithmetic expressions is possible. The representation language L_{LILOG} II includes certain special symbols to allow the creation of arithmetic expressions. These symbols are a special sort symbol NUMBER with the property BOTTOM < NUMBER < TOP, special function symbols +, *, - and special predicate symbols <, >, $\leq$, $\geq$. Furthermore each PROLOG number may be used as a logical constant with the sort NUMBER. Arithmetic expressions are evaluated, immediately using PROLOG. That means, evaluation of artihmetic expressions is only possible, if these are instantiated sufficiently.

L_{LILOG} II also includes a concept of sets, such that, for example, constants can be used to represent sets of objects. The declaration of sorts is extended allowing to declare a symbol to be of a set of a sort. The following special symbols have been incorporated into L_{LILOG} II: empty_set as constant symbol, build_set, union and cardinality as function symbols and is_empty, is_element and is_subset as predicate symbols. To reason about set expressions an ACI-matching algorithm and special algorithms to evaluate ground-expressions will be used.

3.2. Inferencing Tasks

In the first LILOG-Prototype the inference machine was used to derive new facts from the knowledge extracted from the input text (extending the knowledge base) as well as to answer the questions asked to the system (answering questions), (see section 2.2). These tasks are to be performed in the second prototype too.

In the second prototype the question answering capabilities will be improved, taking into account the expectations of a user. In many cases the user expects to receive more information than is explicitly expressed in a question. If, for example, the following question is given:

"Is there a cheap Italien restaurant in Hamburg?",

then it will not be sufficient to answer with "yes". Instead, this answer should be supplemented giving the name of a concrete restaurant. Hence the inference machine has to handle such "yes/no-questions" (see section 2.2.2.2) as "wh-questions" providing a certain kind of over-answering.

Similarly, given the question

"Does Hamburg celebrate the 500th anniversary of its harbour?",

a good answer would not be simple "no" or "don't know", but "no, the 800th anniversary" or at least "I know that Hamburg celebrates the 800th anniversary." Hence if no proof can be found for the given goal, the inference machine should generalize the goal and try to prove it.

Further applications of the knowledge processing are planned for the second prototype. Logical problems, arising during the execution of the components "semantical construction" or "text generation" should be solved using the inference component. Such problems might, for example, be disambiguation and anaphora resolution. To support anaphora resolution the inference machine should determine all reference objects that have certain properties. To reduce ambiguities certain readings can be excluded if an inconsistency is discovered by the knowledge processing.

3.3. The Knowledge Processing Component

This section gives an overview of the knowledge processing component, as it is currently being developed for the second prototype. The knowledge processing component is subdivided into three main modules:

- Selection Strategies,
- Inference Rules and
- Basic Operations.

The module "Selection Strategies" mainly contains the selection strategies and heuristics which depend on the different tasks like extending the knowledge base and answering questions.

The module "Inference Rules" contains the inference rules and is divided into several submodules containing special rules for the treatment of the extensions of predicate logic (e.g. equality, sets), as described in section 3.1. Different calculi may be built up by the various inference rules, depending on the task to be performed by the knowledge processing. E.g., the process of extending the knowledge base uses different resolution rules than the process of answering questions. The following submodules are being developed:

- Resolution-rules,
- Equality-handling,
- Arithmetic,
- Sets.

The third main module contains the basic operations to be used by the inference rules. These are

- Sort-operations,
- Unification and Matching,
- Protocol-operations (explanation, debugging),
- Reason Maintenance,
- Vagueness.

Within this paper we cannot describe each component of the inference machine in detail, but can only give an overview and explain certain aspects briefly.

3.3.1. Selection

Every inferencing task described in section 3.2 will be solved on the basis of a certain calculus which defines the possible operations. However, in general in each step of an inference process several operations are applicable. A selection function is necessary to select a suitable operation from the applicable ones.

The selection modules have to provide functions to

- select some literal and an inference rule,
- check conditions like equality of predicates, difference of signs, values of vagueness,
- unify termlists,
- select one unifier from possibly several most general unifiers (since theory unification is used),
- activate the operation selected,
- manage the other possible operations and unifiers as a basis for subsequent selection processes.

The selection functions of the second prototype will be more complex than those of the first prototype. On the one side there are stronger requirements for the inferencing tasks, and on the other side a more powerful knowledge representation language is used, which is not based on Horn logic, but on full first order predicate logic. Therefore, it is reasonable not to rely on the use of a single search strategy for backward and forward chaining like in the first prototype. Specific selection functions for every inferencing task are required instead.

However, the experience has shown so far, that in most cases knowledge will naturally be modeled in the form of horn clauses. This fact will be exploited, preferring the usual forward and backward chaining strategies for the extension of the knowledge base and answering questions, respectively. Other resolution steps are not excluded, but rather executed when necessary, e.g., when disjunctions or negated literals are involved. Heuristics and the control information specified with the special control language are used to guide the search for a proof, i.e., to select an appropriate operation in each step.

3.3.2. Inference Rules

Knowledge Processing is mainly based on resolution and factorization, which are known to build a sound and complete calculus for first order predicate logic. Besides a general resolution inference rule, special inference rules are used too, in order to support forward and backward chaining between horn clauses. These special inference rules work similarly to the first prototype (see section 2.3).

The treatment of sort formulas will be improved. Order sorted unification and theory resolution as implemented in the first prototype are incomplete for an order-sorted logic with sort literals as pointed out in (Bollinger et al. 1988). For our new formalism we want to implement a complete deduction system based on ideas of the OSSP system which we have developed, and for which we have proven completeness (Müller 1989). This reasoning system simplifies the approach Cohn used in his LLAMA system (Cohn 1987).

The incompleteness of our first approach stems from the following fact.: with order-sorted unification it is not possible to unify a variable X of sort S with a term t of sort S', if $S' \leq S$ is not true. In an order-sorted logic with sort literals, however, the declared sort S' of t may be specialized by a formula of the form p -> S''(t), such that $S'' < S'$ and $S'' \leq S$. Now, if we are able to prove S''(t) (e.g., by proving the premise p of the implication) it should be possible to unify t with X.

This example shows that order-sorted unification has to be extended by a mechanism which takes into account sort information which is expressed by means of sort literals. In OSSP this is done by extending the notion of well-sortedness by allowing more general terms to be substituted for variables than in order-sorted unification. Well-sortedness in the usual sense is implicitly guaranteed by expressing the deviation from being well-sorted by means of conjunctions of respective sort literals (for a detailed description of the OSSP reasoning system see (Müller 1989), the same technique is also described in (Beierle et al. 1989)).

LLILOG II contains the equality relation as a special predicate, which has to be considered by the calculus. Depending on the actual knowledge base, different principles will be used to handle equality: the RUE-calculus of Digricoli (Digricoli 1979) and the paramodulation method (Robinson, Wos 1969).

The RUE-calculus (Resolution by Unification and Equality) is a generalization of resolution incorporating equality. Whereas the usual resolution rule is only applicable if the corresponding subterms of the regarded literals are unifiable, this condition is omitted for RUE-resolution. The literals regarded only must have the same predicate and different sign. If there are corresponding pairs of subterms which are not unifiable, negated equations for these pairs of terms are generated and added to the resolvent to be new subgoals, which still are to be solved.

RUE-resolution should only be applied if the knowledge base contains many equations. If there are only few equations, then there is a high probability for the unsolvability of most of the subproblems. The number of unsolved inequations increases and the search space explodes, since the desired filtering usually gained by unification does not happen.

Another approach to handle equality is paramodulation: Terms are replaced by equal terms. Paramodulation produces huge search spaces if there are many equations in the knowledge base. Many variants of the given formulas might be created without any goal, most of these formulas will never be used.

Hence paramodulation as well as RUE-resolution will be used to handle equality. Depending on the relative size of the set of equations in the knowledge base a calculus based on resolution, factorization and paramodulation or a calculus based on resolution with unification and equality (RUE) is used.

3.3.3. Basic Operations

The different selection functions and inference rules use certain basic operations, which are divided into several submoduls.

Since unification is the central operation for most of the inference rules, appropriate unification and matching (one side unification) algorithms are required. Besides standard unification algorithms, we will also use theory unification (commutativity, associativity) and theory matching (commutativity, associativity, idempotence).

Furthermore the unification and matching algorithms have to consider sort conditions. Unification will be performed by an extension of order-sorted unification. The necessary sort operations are testing the subsort relation (i.e. subsumption relation), computation of greatest lower bounds, and testing, if greatest lower bounds are equivalent to BOTTOM.

Since we do not represent the complete sort hierarchy internally, subsort testing and the computation of greatest lower bounds can not be done by just looking into the sort hierarchy.

In section 3.1.1 we have shown that each subsumption relationship $se \le se'$ can be expressed by the corresponding constraint $(se$ and $\neg se') = $ BOTTOM. Thus for the subsumption test it is sufficient to have an algorithm which for any given sort expression tests if it is equivalent to BOTTOM, i.e., if it is inconsistent.

The computation of the greatest lower bound of two given sort expressions may be done by simply applying the **and** constructor, since the semantics for a sort signature requires that a sort expression built by **and** has to be interpreted by the respective intersection. But in order to test if the greatest lower bound is equivalent to BOTTOM, we again have to test inconsistency for the respective sort expression.

4. Conclusion

In the first prototype we contended ourselves to build up a system able to construct a semantic representation of a text and to answer some rather simple questions about it. For the second prototype our objectives have changed. It is well known that a lot of problems in text understanding are interrelated, e.g. anaphora resolution can not be performed in a satisfactory way without having access to common sense knowledge. Our intention is therefore to use the second prototype also as a research tool for studying these relationship and carrying out experiments to compare different approaches.

For the knowledge processing this requires a high frexibility for reacting on new demands. We hope too meet this requirement by the modular architecture of the knowledge processing component, its enhanced reasoning capabilities and the development of a control language which allows the specification of new inferencing tasks as well as new problem solving strategies. But as first running version of the second prototype is planned for the middle of 1990, experimental results are not yet available.

References

Beierle, C., Dörre, J., Pletat, U., Schmitt, P. H., Studer, R. (1988): "The Knowledge Representation Language L$_{LILOG}$", LILOG-Report 41, IBM Germany, Stuttgart

Beierle, C., Hedtstück, U., Pletat, U., Siekmann, J. (1989): "An Order Sorted Predicate Logic With Closely Coupeled Taxonomic Information", IWBS Report, IBM Germany, Stuttgart

Bollinger, T., Hedtstück, U., Rollinger, C.-R. (1988): "Reasoning in Text Understanding, Knowledge Processing in the LILOG-Prototype". LILOG-Report 49, IBM Germany, Stuttgart

Brachman, R. J., Schmolze, J. G., (1985): "An Overview of the KL-ONE Knowledge Representation System", Cognitive Science 9 (2), 171-216, 1985.

Cohn, A. G. (1987): "A More Expressive Formulation of Many Sorted Logic", Journal of Automated Reasoning 3, 113-200.

Digricoli, V.J. (1979): "Resolution by Unification and Equality". Proc. 4th Workshop on Automated Deduction, Texas

Kamp, H. (1981): "A Theory of Truth and Semantic Representation", in: J. A. Groenendijk et al. (eds.): "Formal Methods in the Study of Natural Language", Vol. 1, Amsterdam

Khenkhar, M.N. (1988): "Vorüberlegungen zur depiktionalen Repräsentation räumlichen Wissens", LILOG-Report 19, IBM Germany, Stuttgart

Müller, B. (1989): "Ein Resolutionsskalkül für ordnungssortierte Prädikatenlogik mit Sortenliteralen", Diplomarbeit, Institut für Informatik, Universität Stuttgart

Pletat, U., v.Luck, K. (1989): "Knowledge Representation in LILOG", to appear in: Bläsius, K.-H., Hedtstück, U., Rollinger, C.-R. (eds.): "Sorts and Typs in Artifical Intelligence", Lecture Notes in Computer Science, Springer Verlag, Heidelberg

Robinson, G., Wos L. (1969): "Paramodulation and TP in First Order Theories with Equality", Machine Intelligence 4, 135-150

Schmidt-Schauss, M. (1988): "Computational Aspects of an Order-Sorted Logic with Term Declaration", Dissertation, Universität Kaiserslautern

Siekmann, J. (1989): "Unification Theory", Journal of Symbolic Computation, Vol. 7, Special Issue on Unification Theory

Stickel, M. E. (1985): "Automated Deduction by Theory Resolution", Journal of Automated Reasoning 1(4), 1985, 333-355.

Wachsmuth, I. (1987): "On Structering Domain Specific Knowledge", LILOG-Report 12 , IBM Germany, Stuttgart

Walther, C. (1987): "A Many-Sorted Calculus Based on Resolution and Paramodulation", in Research Notes in Artificial Intelligence, Pitman, London, and Morgan Kaufmann, Los Altos, Califorina

Indexicality and Representation

Peter Bosch
Institute for Knowledge Based Systems
IBM Germany Scientific Center
Schloßstr. 70, D-7000 Stuttgart 1
bosch@dsØlilog.bitnet

Abstract

In the vast majority of cases linguistic comprehension does not care much about what there is in the world and what state the world is in. Comprehension is often satisfied with satisfaction: you have understood somebody's statement once you have set up a model that satisfies the statement.

But occasionally reality matters. When Albert confesses to Bob: "I'm having an affair with your wife", and Bob's comprehension goes no further than constructing a model that satisfies Albert's statement, i.e. a model in which a speaker is having an affair with his listener's wife, then there is nothing that could make Albert nervous or that could make Bob jealous. Merely constructing a model does not link the statement to the world.

But how do we model this step from a model to real life, from "I" or "you" to a real person ? A proposal sketched near the end of this paper is that - despite first appearance - we don't and we needn't.

1. Introduction

It would seem obvious that there is an important difference between sentences like (1) and sentences like (2). The truth of (1) is independent of when and by whom it is uttered, whereas the truth of (2) depends crucially on who utters it and when. (The examples are Bar-Hillel's (1954:359)).

(1) Ice floats on water.
(2) I am hungry.

When our interest is in relations of logical consequence, cases like (2) are an unpleasant complication, certainly if the definition of logical consequence we want to give is a formal definition, and hence cannot rely on anything but the formal properties of those objects between which relations of logical consequence hold. English sentences that contain indexical expressions, like sentence (2), thus pose a prima facie difficulty. Indexical expressions, like "I", "you", "here", "now", etc. change their reference according to the situation in which they are used. Accordingly, also the truth and falsity of sentences that contain indexicals vary, depending on properties of the situation in which they are uttered, and hence such sentences are not suitable terms for relations of logical consequence.

But what are suitable terms ? When what we are after is a semantics for English sentences then what we need is formal objects that stand in a regular relation to English sentences (at least one such object for each sentence) and that are invariant with respect to relations of logical consequence, as sentences like (2) plainly are not.

The problem at hand is not ambiguity, even though also ambiguities may be, and frequently are, resolved by consideration of the relevant context. By an ambiguous sentence I understand a sentence that is associated with a small and manageable number of different sets of truth conditions. Since ambiguity is due either to the occurrence of an ambiguous word or to a structural equivocation, it can be dealt with by fairly simple disambiguation techniques.

Ambiguous words, or so I shall assume for the purposes of this argument, are listed completely in the lexicon, and their different senses are formally distinguished by different subscripts added to their spelling. And sentences may be regarded as pairs of a sequence of characters (including subscripts, blanks, and punctuation marks) and a structural description. Indeed, the disambiguated sentences we now have are no longer strictly speaking English sentences. But they are formal objects, they are formally distinct from each other, and there is at least one of them for each English sentence.

But such disambiguation techniques, however adequate for ordinary ambiguity, would seem insufficient for context dependence proper. First, there is good reason to believe (cf. e.g. Ziff 1973, Travis 1977, Clark and Clark 1979, Bosch 1983 Ch.3, 1985a) that the number of contextually different word senses is potentially infinite or, at any rate, forbiddingly large, and prevents any form of listing.

Second, and independently: even if the number of word senses were manageable, the variation in reference we find with indexical expressions certainly is not. - So even if the differences and connections between run-of-the-mill ambiguity and context dependence are not clear in all respects, we have here at least a couple of technical reasons not to attempt a treatment of indexicality in terms of established disambiguation techniques.

But we may still try to keep our account of indexicality parallel to disambiguation if, in order to find objects that are invariant under changes of utterance context, we move from English sentences (or, better, disambiguated English sentences) to pairs of a sentence and a context of utterance. These new objects clearly satisfy the requirement of being invariant terms for relations of logical consequence in a way that disambiguated sentences are not. But there still remains one point of worry: can these new objects be construed as formal objects ? Can each two such objects be distinguished on purely formal grounds ?

For convenience of expression I shall refer to these pairs of a disambiguated sentence and a context as "C-propositions" - "propositions", because they seem to be similar in some respects to what others mean by this term, and the "C" as a warning that the term is only used for convenience and no more should be read into it than what follows from our explicit statements. - One may note that C-propositions are particularly close to Stalnaker's 'propositions' and to Kaplan's sentence 'contents' (cf. Stalnaker 1972, Kaplan 1978, Bosch 1982). But I am not interested here in the peculiarities of these or other specific proposals and therefore prefer the non-committal term "C-proposition".

Again the question: can C-propositions be conctrued as formal objects ? Is there a way of distinguishing formally between each two C-propositions ? - We saw above that sentences cause no problem as formal objects. A sentence may be construed as a sequence of characters; and each character may be identified with a set of inscriptions (cf. Quine 1960:195). - But how can contexts of utterance be construed in a similarly unobjectionable fashion ? To this question I shall attend in a moment. But first I must broaden the issue a little.

2. Psychological Considerations

Our interest in the notion of logical consequence with respect to the sentences of a natural language is due to the assumption that this notion plays a central role in an explication of the semantics of a natural language: what follows from the truth of a sentence S tells us a great deal about what S means, at least what S means with respect to, or in terms of, other sentences. Since there is too much fluctuation with respect to logical consequence relations, and hence with respect to the meaning (in this specific sense) of English sentences, we must look for entities that are stable in this respect, for instance C-propositions. C-propositions may thus be regarded as meanings of disambiguated English sentences relative to a context, in the sense that they are equivalent to that part of sentence meaning that is invariant in relations of logical consequence, i.e. that portion of meaning that matters for the preservation of truth.

Understanding an utterance of an English sentence would then, according to the above considerations, be the same as grasping what C-proposition the sentence amounts to when uttered on the particular occasion in question. If this connection were non-existent, the investigation of relations of logical consequence would be irrelevant to an explication of linguistic comprehension.

Now there is the following difficulty. C-propositions are pairs of a sentence and a context. And we require that both sentence and context are formal objects. But do human speakers, when they understand a sentence in a context, avail themselves of anything that would amount to a formal representation of the context ? - In practice this requirement would mean that speakers should be able to 'transcend' their utterance contexts, as it were, to make explicit for themselves what aspects or features of the context are influencing their understanding of the sentence, or to paraphrase each indexical sentence by an equivalent non-indexical sentence. - But ordinary comprehension processes, or so it seems, do not care about such a requirement.

Consider a situation where both speaker and listener neither know the time, nor the place at which they are - two Rip van Winkles, if you like, who not only have slept for an indefinite period of time but who have also been sleep-walking and find themselves in an unknown place . They will, in such a situation, still use indexical expressions that depend on their own current spatio-temporal position. And not only is the functioning of these expressions entirely unproblematic here, but it is - given that neither of them know the place or time - the only way available to them for talking about their spatio-temporal position. - They could not translate their indexical sentences into non-indexical ones, have no way of 'transcending' their situation, and have no representation of their utterance context except their own current and indeed very limited one.

Since human speakers ordinarily communicate by means of indexical sentences without any difficulty of a principled nature, the results of their comprehension processes, at least in these cases, cannot be C-propositions, i.e. representations that make context fully explicit and that are, in this sense, context independent, or non-indexical, representations. The mental representations that result from human comprehension plainly do not contain any information the listener does not have, and if he has no context-independent way of describing, say, the place or the time of the communication situation, some equally indexical equivalent of "here" and "now" will inevitably form part of his mental representation of what he has understood.

In brief, then, the argument runs as follows: relations of logical consequence require terms that are invariant with respect to context; such terms (C-propositions) cannot play a decisive role in comprehension processes, because natural language communication does not require the availability of context-independent representations. Hence: as far as the human processes of linguistic comprehension are concerned, relations of logical consequence cannot play a decisive role either.

3. Two Problems

I have sketched two problems. The first of a general philosophical nature and concerned with the very possibility of a formal reconstruction of context, and the second of a psychological kind and concerned with the role of explicit context representation in human comprehension. While the second problem is taken very serious, by at least some semanticists (cf. Partee 1980 and references there), the first, philosophical, problem has been given next to no attention. The general position seems still very close to that of Bar-Hillel (1954:369) in the very early days of the semantics of indexicality:

"Not in every actual communicative situation could every indexical sentence be replaced, without loss of information, by a non-indexical sentence; but there is, on the other hand, no indexical sentence which could not be replaced by a non-indexical sentence, without loss of information, in some suitable communicative situation".

While Bar-Hillel acknowledges our second problem, he takes a traditional logical empiricist stand with respect to the first. He explicitly endorses Carnap's view that "non-indexical languages are sufficient for the formulation of any body of knowledge" (1954:367).

For the remainder of this paper I shall be concerned with this notion of non-indexical representation. I shall say no more on the psychological question. That, in fact, in human comprehension processes C-propositions and their like do not play a significant role is something I take for granted. But this does not imply that C-propositions are thereby uninteresting for the semanticist. They may well play their part in a philosophical model for human understanding. And a model as such - philosophical or in the empirical sciences - may be adequate, in the sense that it helps us understand certain matters better, even though some parts of the model may stand in no clear relation of correspondence to anything in the object domain which the model is a model of.

4. Non-indexical representation

In order to make precise sense of the notion of a C-proposition along the lines suggested above, and for the purposes suggested above, we need not only a formal notion of a (disambiguated) sentence, but we require an equally formal representation of contexts of utterance.

This requirement may also be formulated by saying that we need something like a type-token distinction for contexts and, with it, a formal definition of the notion of a context type. We are surely not interested, neither as speakers of a language nor as semanticists, in each and every possible difference that there may be between any two actual speech situations. Most differences between real contexts of utterance do not matter for the C-propositions resulting from those contexts, i.e. they could not influence logical consequence. - And this is just as well, because contexts, like any other concrete entities, cannot have complete and finite descriptions. If each and every difference were to matter, the entire enterprise would be doomed to fail from the very start. - What remains troublesome though is the fact that it is not always the same features of real utterance situations that matter for the C-propositions expressed in those situations. If, for instance, I say "I love you", it is usually taken to be sort of important who I am speaking to, but place and time, or climatic conditions and the rheumatism of my neighbour, for that matter, don't seem to contribute much. But when I say "I've never been here before", time and place of utterance become important, while we can't care less about who I am speaking to. And, eventually, also climatic conditions and my neighbour's rheumatism may occasionally play their part, for instance for the understanding of an utterance of "He's always bad in this weather".

This fluctuation in what is and what is not relevant for whether two contexts count as the same would not be so much of a problem if, at least, we had some assurance that the number of parameters that are relevant for the identification of context types would be limited. This, however, is not the case. Lewis (1981:87) acknowledges the difficulty, and so does Cresswell (1973:111f). Cresswell seems convinced, though, that the sentence at issue may help select those parameters that are relevant for the individuation of contexts with respect to that sentence. As long as there are explicit indications in the sentence as to the relevant parameters, such as tensed verbs, adverbs of place and time, or demonstrative pronouns, Cresswell is certainly right. But how is the sentence going to tell us, for instance, that the reading of one of its predicates varies crucially with the context, like e.g. "hexagonal" in the sentence "France is hexagonal" (cf. Lewis 1981) ? Two contexts of utterance that differ along the lines along which typical geographical and typical geometrical contexts differ would in this case have to count as different contexts. If one of the preceding sentences is "Italy is sort of boot-shaped", the corresponding reading of "hexagonal" should make our sentence come out true. But if one of the preceding sentences is "Geometrical shapes are ever so much simpler than shapes in geography", our sentence is less likely to count as true (cf. also Clark and Clark (1979) for a discussion of cases that are probably far more troublesome). - The problem is not that the shape of France has changed, but that the two contexts of utterance must count as different for the purposes of interpreting our sentence.

Or what if the interpretation of a sentence varies with the context of utterance, without there being any particular expression in the sentence that would be responsible ? Take a sentence like "There is coffee in the hall", either as an announcement of a coffee break during a conference, or as an English translation of the triumphant statement of a 19th Century Prussian tax officer ("Kaffeeschnüffler" - as they were called) who has just discovered evidence, in the form of two coffee beans under a rug, of what was probably illegal and untaxed imports. The sentence can hardly be defended as true in the first utterance situation if the fact of the matter consists of those two coffee beans under the rug, whereas no-one would convince our tax man that two beans should be insufficient to make his statement true. - Examples of this kind, where the interpretation of the sentence is clearly context dependent but does not seem to hinge upon any particular expressions in the sentence are easily multiplied (cf. Ziff 1973, Bosch 1985a). And even if we force the responsibility for the context dependence of such sntences on particular words, in the case at hand upon "there is" - perhaps already compositionally limited by its combination with "coffee" - I still fail to see how this could significantly limit the number and kind of parameters that would enter the truth conditions of our sentence.

Hopes for a formal definition of the notion of context type are not flying high when we think of such cases. - And why maintain hope artificially where there is none ? I think we can be more radical. Let me propose the following argument. Cresswell is clearly right when he says that the identity of contexts of utterance is influenced by the sentences whose contexts they are. But this accounts only for one set of factors. Whether or not two real contexts of utterance may count as identical, i.e. as of the same context type, also depends on preceding sentences (most obviously, but not only, in cases where the sentence under consideration contains anaphoric expressions that are coreferential with expressions in preceding sentences) and - via them (or directly) - on t h e i r contexts, plus, of course, on any amount of situational context not mediated via sentences. How then are we to give a formal definition of context types if the semantically relevant context type depends itself on the consideration of contexts ? All we get is recursion.

If this argument is correct, we may wonder about the roots of the identity problem for context types. There seemed to be no such problem with regard to sentence types, or types of disambiguated sentences. And there are infinitely many sentence types just as there are infinitely many context types. How and why are contexts different from sentences ?

The crucial difference is that the set of sentences of a natural language is finitely differentiated (cf. Goodman 1969:135f), whereas the set of utterance contexts is not.

> For each two sentence types, S_i and S_j, and each sentence token s that does not actually count as a token of both types, it is at least theoretically possible to decide either that s is not an instance of S_i or that s is not an instance of S_j.

There is, however, no such assurance for contexts of utterance. There is no such thing as a fixed set of context types in the sense as there is a set of sentence types; in other words: there is no way of knowing what may count as a relevant difference between two context tokens. Or, if we already assume a parametrization of contexts, i.e. a (possibly infinite) list of all those features that may, at least occasionally, count as relevant distinctions between contexts, then there is still no way of knowing whether or not the presence or absence of a particular feature will matter for the identity of any particular context on a given occasion. One of the causes for this problem is that a particular feature of a particular context may count as relevant when the context is approached from one particular preceding context and may not count as relevant when the context is approached from a different context. - And this consideration obviously invites recursion.

At this point the theoretician may be tempted to look to empirical constraints for help: are there not limitations to human perception that could rule out certain differences as irrelevant, thus putting a stop to recursion ? In a sense, yes. There are fairly clear physiological limitations. But no such limitation can have an absolute character, because perception is an intelligent process, far from purely passively receptive. Consider the following Gedankenexperiment, just as an illustration of this more general point by means of a comparatively simple case from visual perception.

Suppose we have a section from the colour spectrum, which we cut into a few thousand or more narrow strips so that we are unable to distinguish the colour of any two neighbouring strips, say A and B in Fig.1. In other words, if our context consists of A and B only, A and B are indistinguishable. Now we add to our context B's other neighbour, C. And suppose that B and C are as indistinguishable from each other as are A and B, while A and C are distinguishable. (I suppose it is reasonable to assume that we can find a way of cutting up the spectrum so that this is actually the case). Now, from what we have observed, indistinguishability of A and B and of B and C, plus distinguishability of A and C, the reasonable conclusion is that A, B, and C must all be different in colour. In other words, in the first context (which consists of A and B only) A and B are indistinguishable, and in the new context, with C added, A and B are distinguishable. The point is not just the familiar one that perception, even here on the very simple level of perception of colour differences, is partly a matter of inference. The more important aspect of our thought experiment is that even below the level of visible difference empirical identity relations may change when the set of contrasting objects, i.e. the context, is changed.

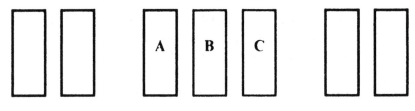

Fig.1

Let me generalize this case. Suppose we have a context, c1, in which two objects, a1 and a2, count as the same thing in the sense of being two tokens of the same type. Either because they are, as in the case just considered, actually indistinguishable in c1, or because, in c1, we cannot, or have no reason to, attach any significance to any differences between them. Both a1 and a2 then are taken as tokens of one type, say the type A.

Now suppose the context c1 is changed to a context c2 by the addition of further objects, among them a3, which is not of type A, but is seen as a token of a further type, A'. It is quite conceivable now that, in c2, where both A and A' are relevant, we would regard a2 as of A', as well as of A, whereas a1 clearly does not form an instance of A'.

Consider for instance a context in which the two objects in Fig.2 count as two instances of the same type, differing only in orientation (which is discounted in the context at hand as irrelevant). This is a natural view in contexts where objects can be turned around, or mirrored, with no restrictions. Now consider Fig.3. The addition of the new, third object, may invite an overall view of the three objects as, say, printing characters. And for printing characters, obviously, orientation matters, and the identity between our first two object has gone. The second and third object however, may now perhaps both be regarded as instantiations of logical 'and'-signs, whereas the the first would count as an 'or'-sign. The first one is then the odd one out, while the context allows us to view the second and third as identical. Still, if we resist the interpretation of all three objects as printing characters and refuse to count orientation as relevant, then the original identity can be maintained and the third object is the odd one out.

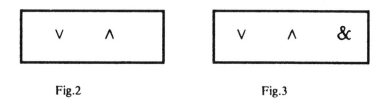

Fig.2   Fig.3

Since perceptual or empirical identity in general can only rest on the irrelevance, or, in the extreme case, the absence, of perceptible difference, i.e. on indistinguishability, and since both relevance and distinguishability depend on the contrasts that are available in the context in which the comparison is made, this means that there is no context-independent way of identifying, and therefore, in the more general case, of classifying anything.

To the extent that contexts are objects like any others, these considerations of identity and classification also apply to contexts. In practice, this can mean either that the identity of a context from one occasion to the other, or the classification of a context as a particular kind of context, depends - in theory - on some wider context within which the question of identity of contexts arises. Or it may mean that identity depends on a preceding context from which we enter a particular new context that must be classified.

Now, plainly, the upshot of all this is that identity and classification of contexts, as it is obviously crucial for linguistic comprehension - i.e. for the identity of sentence-context pairs (C-propositions) - is possible only with respect to some context that is, at least for the time being, taken as primitive. And this is nothing but a generalization of Karl Bühler's (1978:107) notion of an "I-Here-Now"-Origo, the origin of a speakers momentary system of cognitive coordinates.

Let me stress that the latter point is not first and foremost a matter of any particularly human limitations of perception, but is a more general matter of the partiality of experience. It is indeed possible to define identity proper on the basis of indistinguishability:

For any x and any y: x is identical to y iff
there is no z are such that z is indistinguishable from x but not from y.

For any partial experience (and thus all human experience), however, this definition is obviously unavailable.

5. Formal semantics

Formal semantics of natural language is concerned with an explication of semantic notions with respect to natural language. On the most general level, the goal is a characterization of truth and logical consequence for natural languages. This task, according to our argument up to now, crucially involves a formal reconstruction of the notion of a context.

However, as we saw, this notion poses insurmountable problems for a formal reconstruction and thereby endangers the whole enterprise: no formal reconstruction of context, hence no C-propositions, hence no characterization of truth and logical consequence for natural languages - and hence no formal semantics of natural language.

But we are not quite that far yet. In the above exposition, I have followed a traditional - and as now appears, self-destructive - line of thinking in formal semantics, even at points, where there is no need to do so. Most importantly, I have uncritically accepted the implicit notion that relations of inference, as they are essential for natural language semantics, are relations of logical consequence. But is this really so? Would it not be sufficient to have inferences that operate reliably just within one particular context, plus, and this is of course the same, inferences that allow us the transition from one context to the other, provided there is a particular third context within which the first two contexts can be related to each other ? In each particular case, we have then one context, the current context at hand, that is taken as primitive and that provides the required standards of identity for other contexts.

What we want then, instead, or rather on top, of a theory of logical consequence, is a theory of contextual inferencing. Contextual inferencing does not require non-indexical representations, because context is not allowed to vary. It does not matter who the referent of a particular token of "you" is (we don't need a context-independent representation of its referent) as long as we have a mapping from "you" into the particular context within which the reference must be understood. It may happen that this is a mapping onto the same referent that, in this other context, would be referred to as "I". - And what more does the listener need than the information that it is him who is being adressed (never mind a context-independent representation of his identity) ?

Sure enough, such a theory of contextual inference is much less powerful than a theory of logical truth. Notions like logical necessity and possibility cannot be reconstructed in terms of contextual inference. Instead we get epistemological possibility and necessity. And this is probably all we need for natural language comprehension.

6. The philosophy of it

Near the beginning of this paper I assumed, following the tradition of logical grammar, that the explication of relations of logical consequence between sentences is a core area of semantics. The instability of actual sentences of natural language with respect to truth, however, forced us to look for other terms than sentences as the terms of consequence relations. Such terms would need to be formal objects, like sentences, but they have to be insensitive, in their truth conditions, to any changes of context. It turned out that no such context-independent, or non-indexical, representations can be constructed, due to the impossibility of a formal representation of context.

Does this mean that, once we decide to take indexicality and context serious, relations of logical consequence become irrelevant to the semantics of natural language ? Certainly not. It only means that things get a little more complicated. Relations of logical consequence only hold inside one context and once we transgress context boundaries, we first have to reconstruct the preceding context in terms of the new, then current context before any logical inference can be applied. This reconstruction however is not itself a matter of pure logic, but makes use of non-logical rules and axioms, many of them with a default character. The reconstruction of one context from the point of view of another as a whole is not monotonous or straightforwardly incremental. In brief: contextual inferencing comes much closer to the processes that are investigated in AI under the lable "reasoning" than they are to the comparatively simple uses of logic we are familiar with from the tradition of formal semantics of natural language. - The morale of my argument then is not that logic is irrelevant or less relevant to the semantics of natural language, but that the role of logic in the semantics of natural language is a more intricate one than we are inclined to tell our first year students, and than we happily admit to ourselves. - This of course is nothing new at all, but a mere triviality. Having understood the point, the real work is in working out the details. In this respect, the lesson from the above is that natural language semantics and reasoning (in the AI sense) have a good deal of their task in common.

REFERENCES

Bar-Hillel, Yehoshua (1954): "Indexical Expressions". Mind 63:359-379

Bosch, Peter (1982): "The Role of Propositions in Natural Language Semantics". In: W. Leinfellner e.a. (eds.): Language and Ontology. Proceedings of the 6th Wittgenstein Symposium. Hölder-Pichler-Tempsky. Vienna.

Bosch, Peter (1983): Agreement and Anaphora. Academic Press. London.

Bosch, Peter (1985a): "Lexical Meaning Contextualized". In: G. Hoppenbrouwers/ P. Seuren / T. Weijters (eds.): Meaning and the Lexicon. Foris. Dordrecht.

Bosch, Peter (1985b): "Propositionen". In: Thomas Ballmer / Roland Posner (eds.): Nach-Chomskysche Linguistik. De Gruyter. Berlin.

Bühler, Karl (1978): Sprachtheorie (first edn. 1934). Ullstein. Berlin.

Clark, Eve V. / Herbert H. Clark (1979): "When Nouns Surface as Verbs". Language 55:767-811

Cresswell, M.J. (1973): Logics and Languages. Methuen. London.

Davidson, Donald / Gilbert Harman (eds.) (1972): Semantics of Natural Language. Reidel. Dordrecht.

Goodman, Nelson (1969): Languages of Art. Oxford University Press. London.

Kanger, Stig / Sven Öhman (eds.) (1980): Philosophy and Grammar. Reidel. Dordrecht.

Kaplan, David (1978): "On the Logic of Demonstratives". Journal of Philosophical Logic 8:81-98.

Lewis, David (1981): "Index, Context, and Content". In: Kanger / Öhmann (eds.)

Partee, Barbara Hall (1980): "Montague Grammar, Mental Representations, and Reality". In: Kanger / Öhmann (eds.)

Quine, W.V. (1960): Word and Object. MIT Press. Cambridge, Mass.

Stalnaker, Robert (1972): "Pragmatics". In: Davidson / Harman (eds.)

Travis, Charles (1977): "What Semantics Is Not". In: P.A.M. Seuren (ed.): Symposium on Semantic Theory. Dept. of Philosophy. University of Nijmegen.

Ziff, Paul (1972): "What Is Said". In: Davidson / Harman (eds.)

Ziff, Paul (1973): "Something About Conceptual Schemes". In: Glenn Pearce / Patrick Maynard (eds.): Conceptual Change. Reidel. Dordrecht.

CONTEXTUALIZATION AND DE-CONTEXTUALIZATION

Östen Dahl
Department of Linguistics, Stockholm University
S-106 91 STOCKHOLM

1. Introduction

The problem I want to discuss[1] is well illustrated by (1), which originates in Partee 1970 and was made famous by being used by Richard Montague 1974:

(1) The temperature is ninety, but it is rising.

The expression *the temperature* does not by itself refer or denote anything (if it is not used generically). Rather, its interpretation involves a function of at least two parameters: the object whose temperature is measured and the point in time at which it is measured. In one of its most common uses, the 'meteorological' one illustrated in (1), one is talking of the temperature at a certain place rather than the temperature of an object. Let us for the time being restrict the discussion to the meteorological use, and thus speak of the time and place parameters of *temperature*. (In the following, I shall try to make a terminological distinction between 'parameter' and 'argument', using the latter as the syntactic counterpart of the former, even if my usage will not necessarily be quite consistent.)

Linguistically, these parameters may be determined in at least four ways, given a sentence containing the words *the temperature*:

(i) by prepositional phrases modifying the NP:

> the temperature in Hamburg at noon

(ii) by prepositional phrases or similar modifiers somewhere else in the sentence:

> In Hamburg, the temperature was twelve degrees at noon

(iii) by information coming from elsewhere in the text, as in:

> This is a report of the weather in Hamburg at 12 o'clock today. The temperature was twelve degrees.

(iv) by features of the speech situation, as when (1) is interpreted to refer to the temperature at the time and place at which it is uttered

The problem to be discussed, then, is the following: Given that the list of possible sources of the time and place parameters of the noun temperature is not unique for that word *temperature* but rather quite typical of a large part of the vocabulary in natural language, how do we render the flexibility with which we switch back and forth between the different alternatives in everyday language when constructing a formal system?

2. The 'classical' treatment

In Montague 1974, what can probably already be called the 'classical' treatment of sentences like (1) was formulated. Similar proposals can be found around the same time in works by e.g. Scott and Lewis. According to this kind of solution, there is no basic syntactic difference between a noun like *temperature* and most other common nouns. The extensions of all noun phrases, like those of all expressions in the language in general, were supposed to be determined relative to what Montague calls 'indices' or with alternative terminology, 'points of reference' or 'discourse coordinates'. Possible differences between noun phrases would be described in terms of whether the function that picked out individuals relative to the points of reference was constant (as in, presumably, *the Earth*) or not (as in *the temperature*).

Montague's main interest in analyzing (1) lay in the problem how to account for the fact that the temperature could at once 'be identical to' a number ('ninety') and 'be rising' — how can a number be rising? His analysis was, as has been pointed out by later researchers (reference), somewhat simplistic even for (1), but it is easily seen that it covers a relatively small part of the possible uses of the word *temperature* in English — basically only the meteorological sense, and only cases where the parameters are derived from the speech situation (time = 'now', place = 'here'). Cases where the parameters are derived from other kinds of discourse or situational information might presumably be accomodated in theories like that of Lewis (1972), whose 'discourse coordinates' are supposed to include things like 'previous discourse coordinate' and 'prominent-objects coordinate'. Parameters derived from sentence or verb phrase level modifiers could presumably be accomodated by treating those modifiers as operators on discourse coordinates. Parameters that directly modify the noun phrase are harder, however. Montague would himself hardly have seen this as a problem: *temperature* could of course be treated as being ambiguous as to its syntactic category, depending on whether it occurs with explicit parameters or not. While such an approach is in principle possible, its attractivity is diminished by the fact that since *temperature* (in its 'meteorological') reading may occur with either of the time and place parameters, with both of them or with none, we need at least four different distinct lexical items only for this reading of *temperature*, and similarly with all the words that behave in this way.

Sometimes the term 'context' is used to refer to the totality of points of reference/discourse coordinates. There is a clear parallel between this concept and the concept of 'environment' as used in computer science, i.e. the ensemble of parameter or variable values that are valid at a certain point in the operation of a computer. However, there are also important differences in the ways discourse coordinates in the theories discussed here and the variables or parameters in computer programming are thought to work. Whereas the latter may usually be changed both as to their names and their values, the former seem at least in Montague's version of 'indexical semantics' to be more or less God-given, i.e. determined by the external situation. It should be pointed out here that since concrete aspects of language use are normally ignored by theories of this type, it may not always be clear to which extent this statement is true.

3. Defining contextualization and de-contextualization

As a first step towards a general solution of the parameter determination problem, I would like to define two types of operations on linguistic expressions: **contextualization** and **de-contextualization**. The effect of these would be to subtract or add, respectively, arguments to predicates (or other expressions), making them at the same time more or less dependent, respectively, on parameters the context/environment.[2]

Suppose for instance that we have a two-place functor 'the temperature at place x at time t' — call it F. F may then be contextualized in several ways. For instance, we may contextualize F at the time parameter-slot with respect to the time of speech. The result would be a one-place functor 'the temperature at place x at speech time' — call it F'. The value of an expression $F(\alpha, \beta)$ depends on the value of α and β — it is independent of when it is uttered. The value of an expression $F'(\alpha)$ on the other hand depends on the value of α and the time when the expression is uttered. We might also contextualize F completely to yield a new zero-place functor F' whose value is entirely dependent on the speech situation — when and where it is uttered.

De-contextualization, the inverse operation to contextualization, is somewhat trickier to formulate. It involves taking a predicate or a functor F whose interpretation depends on some environmental/ contextual parameter π, yielding a new functor F' which is identical to F except having an extra argument which takes the place of π in the interpretation of F'. To do so, one should in principle know exactly how π influences the interpretation of the expression, which seems to necessitate having access to the full semantic description (definition) of F. However, in a programming language like LISP, it is possible to avoid that problem by inventing a local variable with the same name as the environmental parameter and set its value identical to the value of the added argument.

In Appendix A, I give some definitions in LISP that show how this can be done, with a concrete example. We start out with the zero-place function FATHER, which is defined relative to the variable EGO. When the function FATHER is called, it looks for the symbol which is the value of EGO and picks the value of the property GENITOR (the Latin is just in order to keep things separate) on this symbol's property list. We may now decontextualize FATHER, yielding the new one-place function FATHER-OF. Calling this function with e.g. the parameter JOHN results in the temporary substitution of the JOHN as value for EGO and then calling the function FATHER. The returned value then is the value of the property GENITOR on JOHN's property list.

In fact, both contextualization and decontextualization are made use of in programming languages. Take for instance Pascal, in the popular version Turbo Pascal. A specific field of a record variable may be referred to by an expression which is from our point of view a one- place functor (with the parameter written first), e.g. *thisperson.address* which would refer to the field *address* in the record denoted by *thisperson*. However, if within the scope of the *with ... do* operator, the field designator may stand by itself. For instance *with thisperson do writeln(address)* is equivalent to *writeln(thisperson.address)*. In our terminology, *address* has here been contextualized as to become a zero-place identifier. Conversely, in a program consisting of several 'units', when an identifier is declared in more than one unit, it is always the last declaration that is valid. However, by adding a unit name before the identifier, e.g. *dos.exec* one obtains a reference to the identifier as declared in that unit. This is a fairly clear case of de-contextualization, making the 'unit' parameter, which is normally assigned a value automatically, explicit and manipulable.

4. Context operations in natural language

Let us now consider the possible role of contextualization and de-contextualization in natural language.

Consider the following situation: A developmental psycholinguist studying the language of a small child writes in his diary the following sentence:

(2) Johnnie uses the word ['mam:a] to refer to his mother.

What status does the word ['mam:a] have in Johnnie's language? Well, there is no obvious reason not to regard it simply as a proper name of a salient individual in Johnnie's world. But notice that in describing the child's language, the psycholinguist uses an English noun in a clearly relational way — as a functor with one argument: *his mother*. It follows that there is no direct translation relation between the object language expression ['mam:a] and the meta-language expression *mother*, since they belong to different categories, syntactically and semantically. It is well possible that the child at this stage simply is not able to express relational concepts and that his language does not contain anything corresponding to one-place functors. In other words, the object language that the child uses has a poorer set of categories than and is strictly speaking not commensurable with the meta-language of the psycholinguist.

As the child grows, he will sooner or later discover that also other people have mothers, and he will start saying things like 'your mother'. In my view, the best way of describing this process in not by saying that the child has totally changed his view of what *mother* means: rather he has learnt to apply de-contextualization to his concept of 'mother', which in this case presupposes an ability to take someone else's perspective, i.e. putting as it were someone else in his own place, or in the terminology used above, temporarily changing the value of the parameter EGO.

Consider a similar development, this time historical rather than ontogenetic. Noun phrases like 'the sun' and 'the moon' are often given as paradigm examples of individual terms with unique reference, needing no contextual amplification, yet anaphoric cases are of course possible:

(3) Our spaceship landed on Phobos. The moon was quite desolate.

What (3) illustrates is that as astronomy has developed, the words *sun* and *moon* have become de-contextualized so as to mean 'that which would correspond to the sun in another solar system' and 'that which has the same relation to other planets as the moon has to the Earth' respectively. Again, it strikes me as a linguistically inadequate account to say that *sun* and *moon* have changed their basic arity. In about ninety-nine per cent of all contexts, they are still used as quasi-proper names of our own sun and moon, as is witnessed by the fact that you can capitalize them (*the Sun* and *the Moon*). Rather, de-contextualization should be seen here and in other similar cases as a productive process, yielding new possible uses of old expressions.

Evidence for such a view comes from the existence of certain kinds of metaphors. Metaphors regularly arise by transferring a concept from one domain to another — cf. the botanical and genealogical metaphors used in graph theory (*root, leaf, branch, node, mother, daughter*). It is not uncommon for proper names to achieve metaphorical uses in this way. Consider an expression such as *the Hitler of the 19th century*. In interpreting this we must ask the question 'Who would be the person in the 19th century whose role is most similar to that of Hitler in the 20th century?'. This is tantamount to a de-contextualization of the name *Hitler* — we as it were turn it into a one-place predicate by adding a time argument to it. That we can readily understand such a metaphorically used proper name even if have not heard it that particular name used in that way before shows that decontextualization is a psychologically real process.

Incidentally, the choice of the 'default' value of the time parameter is not always quite trivial. If one says *the Napoleon of our times* we most probably interpret that as 'the person in our time that plays the role Napoleon did during his lifetime'. A phrase such as 'the Marx of the 21st century' could be interpreted similarly, but possibly also as 'the person who will have the same import for the 21st century that Marx has had for our century'.

Let us now add a Tarskian twist to the discussion. Talking about the psycholinguist describing Johnnie's use of ['mam:a] above, I said that there is an important difference between ['mam:a] in the object language and *mother* in the (psycholinguist's) meta-language. The connection to Tarski's discussion of definitions of truth is that de- contextualization may sometimes necessitate enriching the language in a way that really makes it another language. Thus, I said that it was possible that Johnny could not express relational concepts at all. In that case, de-contextualization of the word for 'mother' in his language would entail that it changed its expressive power. Thus, it appears that decontextualization, like truth, is something that cannot be defined within the language itself.

At the same time, decontextualization makes it possible to enhance the expressive power of a language without changing it radically. Take again the astronomical examples used above. Suppose we have a language which is adequate for describing the solar system, as we know it. In such a language the word for 'sun' may well be treated as a constant (= a proper name). In the enhanced language, adequate also on the galaxy level, we allow ourselves to de-contextualize the word for 'sun' to be also a one-place function 'sun of S' where S is any solar system. But this does not prevent us from keeping the old 'zero-place' use of the word − as I said above, this will probably still be the most common use of it. The price to be paid is a systematic ambiguity in the lexicon, but the gain is that the old 'solar-system level' language is retained as a subset of the extended 'galaxy' language.

5. Contextualization and de-contextualization in discourse processing

If one agrees that contextualization and de-contextualization do play a role in natural language in general, the question to be asked is to what extent they can throw light on on-line discourse processes. The original title of this paper was 'Missing arguments': in choosing it, I was mainly thinking of the fact that texts/discourses are full of nouns, adjectives and verbs with apparent 'implicit arguments', that is, expressions whose semantic interpretation would have to involve the 'filling in' of some parameter: what Fraurud 1989 calls 'parameter activation'. However, it should be clear from the discussion above that not every situation which might at first sight be seen as involving parameter activation does in fact have to be treated that way. As we saw, even if the child's word ['mam:a] might look like a relational noun, it makes little sense to think that the child himself has to activate some parameter in order to understand a concrete use of it. An adequate theory of discourse processing must thus be able to delimit the situations in which the concept of 'parameter activation' is applicable. However, the choice between using this concept and not using it will sometimes be somewhat arbitrary. This goes in particular for the cases when a parameter is assumed to have a default value. From a processing point of view, it may often be as far-fetched to speak of the 'choice' of a default value as to say that I 'choose' to sleep in my own bed instead of e.g. my neighbor's.

A full treatment of the problem of delimiting parameter activation would demand at least another paper. I shall restrict myself here to a couple of observations.

Looking at 'relational' nouns, i.e. nouns whose interpretation involve the determination of some parameter value, we may observe that the character of the parameter may differ quite considerably. In some cases, the value of the parameter is something quite specific and concrete − e.g. an individual (person or object), as in *wife* or *husband*. In other cases, the value is a set or class of objects (a 'domain'), the delimitation of which may be more or less well-determined (as in *the elite*). In still other cases, the value is rather an abstract entity such as a process, an event or some vaguer concept like 'the situation'. A newspaper headline such as *Smith is the winner* demands for its interpretation that we know what Smith won — it might be a specific contest of some kind but might equally well be something less well-defined, e.g. 'this year's political life'. Lexical items in natural languages may be ambiguous not only with respect to the number of parameters they involve but also with respect to parameter type. The Russian word *nevesta* has at least two English translations: *fiancée* and *bride*. We may note that the two latter differ in that their interpretation (in their most common use) involves a parameter whose value is an individual for *fiancée* and an event (a wedding) for *bride*. When children develop a relational understanding of words like *mother*, they tend to interpret it relative not to a person but rather to a whole family. This leads to their using e.g. *your mother* in speaking to an adult referring to his wife (i.e. the person called *Mother* in his family).

We may see less concrete and specific parameters like the ones discussed in the preceding paragraph as an intermediate stage between a non-relational and a relational interpretation of a word. The less specific the value of a parameter is, the less plausible is the hypothesis that a speaker/listener actually uses a process of parameter activation in interpreting it.

FOOTNOTES

1. The discussion in the present paper is partly a development of some ideas originally presented in Dahl 1975. I have been informed that similar arguments have been put forward by Jon Barwise; regrettably, I have not yet been able to get hold of the text in question.

2. In various descriptions of natural languages, 'valency-changing' operations have been proposed. For instance, the intransitive use of the verb *eat*, as in *John is eating*, might be regarded as being derived from the transitive *eat* by an intransitivizing operation, semantically equivalent to an existential quantification of the object slot of the transitive verb. The operations proposed here are different, however, in that the value of the deleted/added argument is taken from the context/environment as defined in the main text.

REFERENCES

Dahl, Ö. 1975. On Points of Reference. Semantikos 1.45-61.
Fraurud, K. 1989. Towards a Non-Uniform Treatment of Definite NP's in Discourse. In Ö. Dahl and K. Fraurud (eds.), Proceedings of the First Scandinavian Workshop on Text Comprehension. Stockholm: Univ. of Stockholm, Dept. of Linguistics.
Lewis, D. 1972. General Semantics. In D. Davidson & G. Harman (eds.), Semantics of Natural Language. pp. 169-218. Dordrecht: Reidel.
Montague, R. 1974. The Proper Treatment of Quantification in Natural Language. In R. Thomason (ed.), Formal Philosophy: Selected Papers of Richard Montague, p. 247-270. New Haven and London: Yale University Press.
Partee, B. 1970. Opacity, Coreference, and Pronouns. Synthese 21.359-85.

APPENDIX A

EXAMPLES OF LISP DEFINITIONS

```
(DEFUN CONTEXTUALIZE (NLAMBDA (FUNCTION VARIABLE NUM)
 (IF (NULL NUM) (SETQ NUM 0))
 (LIST 'LAMBDA 'LST
(LIST 'APPLY FUNCTION
(LIST 'INSERT-NTH (LIST 'EVAL VARIABLE) 'LST NUM)))))
(DEFUN DE-CONTEXTUALIZE (NLAMBDA (FUNCTION VARIABLE)
(LIST 'LAMBDA 'LST
(LIST 'LET (LIST (LIST VARIABLE (LIST 'CAR 'LST)))
 (LIST 'APPLY FUNCTION (LIST 'CDR 'LST)))))))
(DEFUN FATHER-OF (X) (GET X GENITOR))
(PUTD FATHER (CONTEXTUALIZE FATHER-OF EGO))
(PUTD FATHER-AGAIN (FUNCALL (DE-CONTEXTUALIZE FATHER EGO) X))
```

PRINTOUT OF SAMPLE LISP SESSION USING THE FUNCTIONS DEFINED ABOVE

```
$ (SETQ EGO JOHN)
PELLE
$ (PUT EGO GENITOR BILL)
BILL
$ (PUT BILL GENITOR TOM)
TOM
$ (FATHER-OF EGO)
BILL
$ (FATHER-OF BILL)
TOM
$ (FATHER)
BILL
$ (FATHER-AGAIN EGO)
BILL
$ (FATHER-AGAIN BILL)
TOM
```

Computational Semantics: Steps towards "Intelligent" Text Processing

Jens Erik Fenstad   *Jan Tore Lønning*
Dept. of Mathematics
University of Oslo
P.O. Box 1053, Blindern, 0316 Oslo 3
Norway

The title is a programme. Today we have systems which search in, rearrange and check texts for a number of purposes. These systems are necessary tools in any processing of information. But they have limitations. The underlying algorithms typically operate on the given text as a set of symbols or signs - as a stream of machine codes. But signs are carriers of structure and meaning. And this is a fact that we have to take seriously if we are to develop "true" information processing systems. This is, in particular, urgent if we aim at systems which can analyze and structure information from documents which simultaneously contain text and pictures and where the understanding of the full document depends on the joint and interdependent understanding of the component parts.

This sets a goal for computational semantics, a field of study which lies at the intersection of three disciplines: linguistics, logic and computer science. A text processing system should provide a coherent framework for relating linguistic form and pictorial representation with the semantic content of the document. And the relationship must be algorithmic.

There have been important interactions between pairs of the three disciplines. Formal language theory and computational syntax and morphology are to a large extent the result of a marriage of linguistics and computer science. Natural language semantics became more firmly based when one saw how to apply the technical tools developed for the semantic analysis of logical formalisms to language itself. And proofs and computability have formed a strong link between logic and computer science. A more recent and important trend in "applied logic" is the use of model theory in the study of knowledge representation.

The added dimension of computational semantics is the insistence on relating all three disciplines simultaneously, to move into the non-empty intersection between linguistics, logic and computer science. And this is necessary from a cognitive as well as a knowledge engineering perspective.

We can approach the "common intersection" from different starting points. One - and this is the tradition of computational linguistics - is to start from linguistic form and to move to some representational format which is better suited for further processing. Another - and this should be the perspective of knowledge representation theory - is to start from an analysis of the "real world" structure of the problem before us and to develop its mathematical form or model structure. The logical formalism and the associated proof theory are secondary, being the tools we use to investigate the structure of a class of models. An axiomatization theorem would mean that we have succeeded in characterizing the structural properties of the model within the proof system. And the proof system would be the starting point for extracting algorithms for efficient inference procedures.

Representations derived from linguistic form and a model theory derived from a structural analysis must match in order to have a coherent computational semantics. In the first part of this lecture we follow the approach starting from linguistic form and review some recent theories on how to represent the "logical form" of a text and how to connect this representation with a model theory for many-sorted and partial information. The model theory has a complete proof procedure, which has been used as a starting point for developing efficient algorithms for partial logic. In the second part of the lecture we give an example of the second approach and start by discussing a model theory appropriate for the analysis of plural noun phrases. A logic PL for plurals is introduced and used to investigate the structure of the model. Finally we make some remarks on questions of logical strength and effective proof procedures for the logic PL. The matter is delicate, the logic PL has second order features, but is it necessarily of the full strength of second order logic and hence nonaxiomatizable?

Representations

In the transition from linguistic form to semantic content there are reasons to separate out several stages. In the first stage we pass from "given" form to an intermediate representational level. The given form may be a syntactic structure, but syntax alone is not the only constraint on meaning. The representational level may also encode information from phonology, morphology and the larger context of an utterance. We may also, if e.g. the source is a complex document, have to encode pictorial information at the representational level. In a somewhat suggestive language, we hope that the representational level depicts the "logical" form or structure of the source form. And this need not be a formula or set of formulas in some specific logical language such as higher order intensional logic or some similar system. In fact, a different representational mode has emerged as the preferred form: feature structures or attribute-value systems.

The literature on feature valued systems is vast and we have nothing fresh to contribute in this lecture; for a survey from our point of view, the reader is referred to Fenstad, Langholm and Vestre (1990). But to remind the reader of a few salient features, we recall three examples from linguistics as presented in this survey.

Lexical-Functional Grammar

Our first example is the LFG theory due to J. Bresnan and R. Kaplan, see the collection Bresnan (1982) as well as the lectures of P. Sells (1985). We recall that the **grammar** is determined by a **lexicon** and a simple syntactic structure, the **c-structure**. Information is basically given in the form of equational constraints, and a grammatical analysis leads up to a set of equations, the **f-description,** where the unknown entities are attribute–value structures.

Everyone's example of a c-structure is

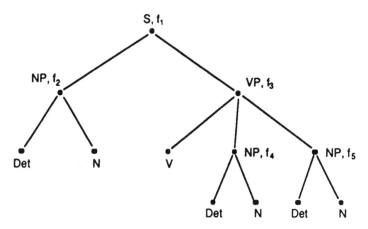

with the associated structural equations

$(f_1\ SUBJ) = f_2$

$f_1 = f_3$

$(f_3\ OBJ) = f_4$

$(f_3\ OBJ2) = f_5.$

We read the meaning of this from the syntax tree and the set of equations: the S-node and the VP-node have a common associated attribute–value structure, $f_1 = f_3$, which has at least three attributes or features, $SUBJ$, OBJ, $OBJ2$, and possibly others. So far we get a representation

$$f_1, f_3 \begin{bmatrix} SUBJ & f_2 \\ OBJ & f_4 \\ OBJ2 & f_5 \\ - & - \\ - & - \end{bmatrix}$$

where the values f_2 ,f_4 ,f_5 are further attribute–value structures, atomic or complex.

Recalling the "standard" example

(1) a girl handed the baby a toy,

the lexicon adds the following lexical constraint equations

$(f_2\ SPEC) = A$ $(f_4\ NUM) = SG$

$(f_2\ NUM) = SG$ $(f_4\ PRED) = \ 'BABY'$

$(f_2\ NUM) = SG$ $(f_5\ SPEC) = A$

$(f_2\ PRED) = \ 'GIRL'$ $(f_5\ NUM) = SG$

$$(f_3 \; TENSE) = PAST \qquad\qquad (f_5 \; NUM) = SG$$
$$(f_4 \; SPEC) = THE \qquad\qquad\quad (f_5 \; PRED) = \;'TOY'$$

$$(f_3 \; PRED) = \;'HAND < (\uparrow \; SUBJ)(\uparrow \; OBJ2)(\uparrow \; OBJ) >'$$

A solution of the combined set of structural and lexical equations can be represented as a feature structure:

$$
\begin{bmatrix}
SUBJ & \begin{bmatrix} SPEC & A \\ NUM & SG \\ PRED & 'GIRL' \end{bmatrix} \\
OBJ & \begin{bmatrix} SPEC & THE \\ NUM & SG \\ PRED & 'BABY' \end{bmatrix} \\
OBJ2 & \begin{bmatrix} SPEC & A \\ NUM & SG \\ PRED & 'TOY' \end{bmatrix} \\
TENSE & PAST \\
PRED & 'HAND < (\uparrow \; SUBJ)(\uparrow OBJ\,2)(\uparrow \; OBJ) >'
\end{bmatrix}
$$

Grammaticality of the given string is not guaranteed by the mere existence of an associated attribute–value structure. A **first condition of grammaticality** is to require that the string has a **valid c-structure** with an associated f-description (i.e. set of equations) that is both **consistent** (i.e. has a solution) and **determinate** (i.e. has a unique minimal solution). Obviously, the standard example satisfies this condition.

But how do we ensure that the final f-structure contains all and only those grammatical roles that the predicate calls for? We call an f-structure **locally complete** iff it contains all the governable grammatical functions that its predicate governs. We call it **complete** iff it and its subsidiary f-structures are locally complete. In a similar way, we call an f-structure **locally coherent** iff all the governable grammatical functions that it contains are governed by a local predicate. We call it **coherent** iff it and its subsidiary f-structures are locally coherent.

A **second condition of grammaticality** now requires that a string is grammatical only if it is assigned a complete and coherent f-structure. Again, the standard example fits the bill.

We used "only if"; even adding the second condition on grammaticality is not enough to weed out all unacceptable strings. We shall return to this problem in connection with our third example.

Situation Schemata

Situation schemata were introduced in Fenstad et al. (1985, 1987) in order to link the grammatical analysis of LFG with the Situation Semantics of Barwise and Perry. As we remarked in the introduction, the scope is wider than this particular linking. In the transition from f-structures to situation schemata we see a shift from grammatical to semantical roles. An example will bring this out. Let our sample sentence be:

(2) John married a girl.

The associated situation schema will be

$$
\begin{bmatrix}
REL & marry \\
ARG.1 & \begin{bmatrix} IND & John \end{bmatrix} \\
ARG.2 & \begin{bmatrix} IND & IND.1 \\ SPEC & A \\ COND & \begin{bmatrix} REL & girl \\ ARG.1 & IND.1 \\ POL & 1 \end{bmatrix} \end{bmatrix} \\
LOC & \begin{bmatrix} IND & IND.2 \\ COND & \begin{bmatrix} REL & \prec \\ ARG.1 & IND.2 \\ ARG.2 & IND.0 \end{bmatrix} \end{bmatrix} \\
POL & 1
\end{bmatrix}
$$

The "mechanics" of how to arrive at the schema from the given string is explained in the book Fenstad et al. (1987); see also the ACAI'87 lectures by Fenstad (1988).

Here we only want to draw attention to two points:

The value of the *ARG.2* feature is basically a notation for a generalized quantifier; hence our representational form is well suited to bring out the essential identity between noun phrases and generalized quantifiers.

The shift from the *TENSE*-attribute of LFG to the *LOC*-attribute of a situation schema allows for a richer "geometric" representation than merely recording the tense of verbs. This is of special importance in analyzing locatives of different kinds. Our next example

(3) John ran to the school,

is meant to illustrate this point. Here the associated schema is

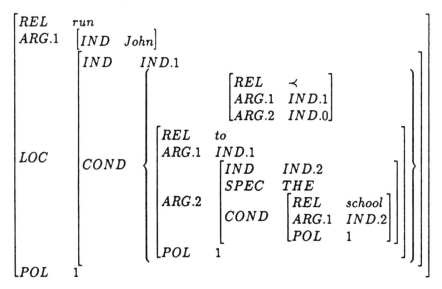

In this analysis $IND.2$ represents an object, *the school*, and $IND.1$ represents a trajectory in space-time which stands in the "to-relation" to (i.e. ends at) the school. An **analysis of locative prepositional phrases** along these lines has been given by E. Colban (1987).

Other, but related questions of semantic representation is discussed by Halvorsen (1988) and in Halvorsen and Kaplan (1988). The format of situation schemata is also being used by R. Johnson, M. Rosner and C. J. Rupp (1990), by H. Dyvik at the University of Bergen, see Dyvik (1988), and by the KOLIBRI project led by D. Metzing at the University of Bielefeld, see e.g. Meier et al. (1988).

Head-Driven Phrase Structure Grammar

HPSG developed historically out of GPSG, but seen today from a somewhat different perspective it represents a "purer" version of an attribute–value formalism than standard LFG. The basic text is C. Pollard and I. A. Sag (1987).

HPSG can been seen as a **theory of signs**:

$$
\begin{bmatrix}
PHON \\[4pt]
SYN \quad
\begin{bmatrix}
LOC \quad
\begin{bmatrix}
HEAD \\
SUBCAT \\
LEX
\end{bmatrix} \\
BIND
\end{bmatrix} \\[4pt]
SEM
\end{bmatrix}
$$

A sign can be **phrasal** or **lexical**, according to whether the LEX feature is unmarked or marked.

The minimal syntactic analysis of LFG is in the HPSG theory encoded directly into the sign or the attribute–value structure by using a special set of features. A sign can at a given level have a feature $DTRS$, daughters, with values $HEAD\text{-}DTR$, $COMP\text{-}DTR$, ... If you are at the sentence level, the path leading from $DTRS$ to $HEAD\text{-}DTR$ will take you down the branch of the corresponding syntax tree in LFG leading from the S-node to the VP-node. And the path leading from $DTRS$ to $COMP\text{-}DTR$ will take you down the branch leading from the S-node to the NP-node. Thus a saturated sign in HPSG will retain a complete copy of every step of the syntactic analysis, whereas in LFG the final f-structure will only display the predicate, the grammatical roles and the possible modifying circumstances.

The **head feature principle** is one of the basic principles of universal grammar in HPSG. It ensures the sharing of certain information between the $HEAD$ attribute of one level and the $HEAD$ attribute of $HEAD\text{-}DTR$ of the next lower level, and corresponds to the "*up-arrow = down-arrow*" equation of LFG.

The **subcategorization principle** is another major principle of universal grammar according to HPSG. A complete sign must be **saturated**. A noun in itself is not saturated, but calls for a determiner to form a full sign, an NP. In the same way a VP subcategorizes for an NP in order to form a full sign, in this case a sign of category S. We see the relationship between this principle and the second condition of grammaticality in LFG, which requires completeness and coherence.

According to HPSG a language is given by an "equation," say

$$ENGLISH = P_1 \wedge \ldots \wedge P_n \wedge P_{n+1} \wedge \ldots \wedge P_{n+m} \wedge (L_1 \vee \ldots L_p \vee R_1 \vee \ldots \vee R_q)$$

where

$P_1, \ldots, P_n$ are the principles of universal grammar, e.g. the head feature principle, the subcategorization principle, etc.

$P_{n+1}, \ldots, P_{n+m}$ are the language-specific constraints.

$L_1, \ldots, L_p$ are the lexical signs.

$R_1, \ldots, R_q$ are the grammar rules.

The meaning of this "equation" is rather straightforward. A sign S is a sign of English iff

- it satisfies all principles of universal grammar, i.e. shares information according to the head feature principle, fills roles correctly according to the subcategorization principle, ... ;

- it satisfies the language specific constraints such as special constituent ordering principles;

- it is either a lexical sign, or satisfies one of the grammar rules of the language.

It is very instructive to compare the syntactic analysis of LFG and the semantic analysis of the Situation Schema approach with the analysis of HPSG. The basic approach of HPSG is to encode everything into the sign. A different approach to attribute–value grammar formalisms, close in spirit to the original LFG analysis, has been given by M. Johnson in his thesis; see Johnson (1988).

Both Pollard and Sag (1987) and Johnson (1988) go beyond the grammatical analysis to a formal analysis; the reader is referred to Fenstad, Langholm and Vestre (1990).

Representations and Interpretations

Representations must interact with the model theory, otherwise there will be no computational semantics. We have briefly concentrated on the representational part of the enterprise; we must now pay attention to matters of interpretation and knowledge representation.

First order logic is a theory of great elegance and depth. But as a knowledge representation tool it has its severe limitations. Recently another "metaphore" has started to emerge: a **conceptual space** is a many-sorted structure where each domain is a separate quality dimension, e.g. colour, weight, space, time, temperature, mass, ... It is always assumed that each quality dimension has some topological or metrical structure.

The notion of conceptual space has roots both in "formal" physics, i.e. in the modelling process in the natural sciences, and in "naive" physics, i.e. in the modelling process in AI and in knowledge representation theory. The importance of the present point of view has been argued by P.Gärdenfors (1987). It is also tied up to the renewed interest in many-sorted logic within AI, see A.Cohn (1987). We have on previous occasions expanded on this point of view and shall not repeat ourselves in this lecture, see Fenstad (1988) and Fenstad, Langholm and Vestre (1990).

In connection with the representational format of situation schemata we have studied a logic of situations and partial information. This is a formal system which is rooted in the situation semantics of Barwise and Perry (1983), see the above surveys or the "basic text", Fenstad, Halvorsen, Langholm and van Benthem (1987). Recently T.Langholm has developed efficient algorithms for this logic, see Langholm (1989). A "geometric" extension is now beeing developed by E.Colban as part of his studies on locative prepositional phrases.

To see in one example how representations and interpretations combine to determine the meaning of an utterance, we recall a simple example. But first we have to remind the reader of some notions from the general theory.

Situations and Basic Facts

A central notion will be that of **basic facts.** These come in several varieties. Typically, a basic fact states that a certain relation holds (or does not hold) among certain individuals at some space–time location. This calls for three sets of primitives: relations (R), locations (L) and individuals (D). We use the format

$$at\ l : r, a_1, \ldots, a_n; 1$$

to express that at the location l in L the relation r in R holds of the individuals $a_1, \ldots, a_n$ in D. Substituting 0 for 1 we obtain the dual fact, that the relation does *not* hold among the given individuals at the specified location. The facts of this first variety we call **located**, since a location is specified. **Unlocated** facts are of the type

$$r, a_1, \ldots, a_n; i$$

where i is 0 or 1. These typically involve a relation of a more permanent kind, such as that of a_1 being the mother of a_2. To such relations time and place are irrelevant, hence they do not enter into the corresponding facts.

Certainly these are not all the varieties of basic facts that can be envisaged. In this exposition we include a third type, to express relations between locations, viz. temporal overlap and temporal precedence. Thus, the location fact

$$l\ overlap\ l'$$

states that the two space–time locations l and l' have a temporal overlap. Such a fact can be useful for analyzing sentences in the present tense.

A **situation** determines (or **contains**) a set of located and/or unlocated facts. We write

$$in\ s : at\ l : r, a_1, \ldots, a_n; 1$$

or

$$in\ s : at\ l : r, a_1, \ldots, a_n; 0$$

if the fact following the first colon is contained in the situation s. Similarly for unlocated facts.

A situation may be **partial** in the sense that many (or most) issues will not be settled either way by the situation, i.e. for several r and $a_1, \ldots, a_n$ the situation s may contain neither of the facts

$$r, a_1, \ldots, a_n; 1 \qquad r, a_1, \ldots, a_n; 0.$$

Similarly for located facts. A situation is thus a kind of restricted, partial model (data base) containing a limited collection of facts.

The choice of basic attributes in a situation schema matches the structure of basic facts

$$
\begin{bmatrix} REL & - \\ ARG.1 & - \\ . & . \\ . & . \\ ARG.n & - \\ LOC & - \\ POL & - \end{bmatrix}
\quad
\begin{bmatrix} REL & - \\ ARG.1 & - \\ . & . \\ . & . \\ ARG.n & - \\ POL & - \end{bmatrix}
\quad
\begin{bmatrix} REL & - \\ ARG.1 & - \\ ARG.2 & - \end{bmatrix}
$$

In the first two matrices, the attributes $REL, ARG.1, ..., ARG.n$ and LOC correspond to the primitives of relations, individuals and locations. POL, abbreviating polarity, takes the value 1 or 0. The values in the matrices are either atomic or embedded attribute–value matrices. The value of the attribute LOC is always complex.

The reader is referred to the book Fenstad et al. (1987) for a complete exposition of the basic theory. The basic reference on Situation Theory is Barwise and Perry (1983). For a current review see the recent collection Barwise (1989).

Meaning and Interpretation

The interpretation of a situation schema in a situation structure is always relative to an "utterance" situation u and a "described" situation s. The utterance situation is further decomposed into a discourse situation d and a speaker's connection c. The former provides information about who the speaker is, who the addressee is, the sentence uttered, and the discourse location in space and time; the latter is a map which fixes the speaker's meaning of lexical items. The **meaning** of a sentence φ is a relation between an utterance situation d, c and a described situation s, and we use the situation schema computable from the syntactic form of φ to express this relation. We write this as

$$d, c[SIT.\varphi]s,$$

where $SIT.\varphi$ is the situation schema computed from φ. We shall give a simple example to illustrate the above concepts. Our sample sentence will be

(4) φ : John is running.

The situation schema of this sentence is

$$SIT.\varphi \begin{bmatrix} REL & run \\ ARG.1 & John \\ LOC & \begin{bmatrix} IND & IND.1 \\ COND & \begin{bmatrix} REL & \circ \\ ARG.1 & IND.1 \\ ARG.2 & IND.0 \end{bmatrix} \end{bmatrix} \\ POL & 1 \end{bmatrix}$$

The atomic values $IND.0$ and $IND.1$ are called **indeterminates**. Let d, c be an utterance situation. A map g defined on the set of indeterminates of $SIT.\varphi.LOC$ with values in the set L of locations is called an **anchor** on $SIT.\varphi.LOC$ relative to d, c if

$g(IND.0) = l_d$, and

$g(IND.1)$ *overlap* l_d,

where l_d is the discourse location determined by d. (Hence the relation symbol $\circ$ is interpreted as *overlap*.) Relative to a described situation s we now have

$$d, c[\![SIT.\varphi]\!]s$$

if and only if there exists an anchor g on $SIT.\varphi.LOC$ relative to d, c such that

$$in\ s : at\ g(IND.1) : c(run), c(John); 1.$$

Observe that the speaker's connection is a map defined on the (morphological) parts of the expression φ and with values in the appropriate domains, i.e. $c(run)$ is an element of R and $c(John)$ is an element of D.

In Fenstad et. al. (1987) we introduced a series of logical systems as tools for the study of the model theory of situations and partial information. The proof theory of these systems of logic can be used as an inference mechanism as e.g. in the question-answer system developed by E. Vestre (see Fenstad, Langholm and Vestre 1990). In Langholm (1989) there is a general study of algorithms for partial logic based upon a novel conception of the Gentzen calculus.

The model structures mentioned so far share one basic feature, they are all individual based. The abstract model structure is in some sense "atomic". But not all properties are atomic, i.e. properties of individuals. Mass nouns such as *gold, water* ... are not primarily seen as properties of individuals. A "correct" analysis requires a different underlying model structure. This has now been developed by a number of authors, see G.Link (1983) and J.T.Lønning (1987a) for some recent contributions and Lønning (1989b) for a discussion of how these analyses may be made computational.

A proper grasp on how to combine the syntax and semantics of comparatives also seems to call for an extension of the standard model structures of first and higher-order logic; see the paper *On the Logical Structure of Comparatives*, by M.Pinkal (1990).

The Logic of Plural Noun Phrases

For the rest of the paper we will study more systematically one phenomenon that goes beyond the individual based model structures and shows the importance of using more structured models: the semantics and logic of plural noun phrases. In particular we will consider the so-called collective readings of such noun phrases. Consider examples like

(5)
  (a) John and Harry
  (b) The boys
  (c) A boy and a girl
  (d) Some boys
  (e) John and some boys
  (f) Two boys and three girls
} bought a boat together.

In sentence (5b), we ascribe a property to the collection *the boys*, and not to each boy. This cannot be represented by a formula in first order logic where the non-logical symbols are the unary predicates *boy* and *boat* and the binary *bought*.

There have been several proposals for the semantics of such constructions. Rather than proposing yet another one, we will describe a unified framework which subsumes the various proposals and which may serve as a workshop for comparing them (Lønning 1989a).

We will make one basic assumption for our format of representation. Objects referred to by plural NP's like *the boys*, *John and Harry* have the same ontological status as objects referred to by singular NP's like *John* and should consequently be represented by terms of the same logical type. This assumption, which is by now widely shared (Blau 1981, Hoeksema 1983, Landman 1988, Link 1983, Massey 1976, Scha 1981), is by us not rooted in assumptions of the type "sets are abstract objects that cannot buy boats". Rather the assumption is grounded in an observation about how natural languages, like English, behave. Disregarding number agreement, English exhibits a nearly uniform behaviour with respect to singular and plural NP's. In particular, a question like *Who bought the boat?* can alternatively be answered with *John, John and Harry* or *the boys*. An approach which treats the two NP's as of different logical types (Bennett 1975) will have problems with explaining this and will easily be led to the confusion that the question is ambiguous, which it is not.

This exemplifies another more basic assumption underlying our approach in this section. A semantic representation ultimately aims at relating a language expression with its meaning in the world. It should interact both with the linguistic expression and with the real world. But as a representation of the information present in the particular linguistic expression, the relationship to the language has priority. We will hence follow the practice from logical model theory and let the semantic entities be determined from the language itself. The entities in our representation will correspond to constructions in English rather than to entities anticipating what the underlying "real-world" semantics will be, i.e. we approach the topic in the spirit of what E. Bach has called "natural language metaphysics".

Eventually, there might be a need for more levels "closer to the world", e.g. when text and pictures interact in the interpretation of a complex document. This would be reminiscent of the relationship between "language-near" tense logic and "world-near" temporal logic (Fenstad et. al. 1987, van Benthem 1982) and between a "language-near" model theory for spatial prepositions and a geometric cartesian description of the space.

More concretely, the formal language, which we here introduce to study the model theory of plural noun phrases, will be first order logic extended with constructions corresponding to the NP's in example (5). We hope that the partial first-order logic of Fenstad et al. (1987) can be similarly extended. Furthermore, we expect that automated deduction procedures for the logic described here can be developed as extensions of similar procedures for first order logic.

The Language PL

The basic idea is that the domain of individuals E shall be extended to a larger domain of objects O, $E \subseteq O$. E shall contain all basic individuals like John and each horse. O shall in addition contain more complex objects like the object that has John and Harry as parts and the object that is constituted by all boys and denoted by *the boys*. We will informally call these objects collections. Consequently, the representation language will contain three types of non-logical symbols:

- Constant symbols, like *john*, *harry*, corresponding to proper nouns in English and denoting members of E. They will be called e-**terms**.

- Unary predicates like *horse*, *boy* corresponding to common nouns in English and denoting subsets of E. They will be called e-**set terms** or **terms of type** $\langle e, t \rangle$.

- n-ary relation symbols for $n = 1, 2, \ldots$, like unary *run*, binary *buy*, corresponding to verbs in English and denoting sets of n-tuples from O. They will be called **terms of type** $\langle o^n, t \rangle$. In particular, terms of type $\langle o, t \rangle$ (unary relations) will be called o-**set terms** or only **set terms**.

The set E is determined to be large enough to contain all denotations of CN's as subsets. Since a verb like *buy* also has more general collections in its extensions, like *John and Harry*, it denotes a subset of $O \times O$ and not a subset of $E \times E$. But observe that so far the system is neutral with respect to whether E and O are different sets; they might be equal.

The language contains in addition a (nearly) standard machinery:

- A countable set of e-**variables**: $x_1, x_2, \ldots$ which take values in E. They are e-**terms**.

- A countable set of o-**variables** (or simply variables): $y_1, y_2, \ldots$ which take values in O and which will be called o-**terms** or only **terms**.

- Two logical e-**quantifiers** *Some* and *Every*.

- Whenever R is of type $\langle o^n, t \rangle$ and each $t_1, \ldots, t_n$ is of type e or o then $R(t_1, \ldots, t_n)$ is a **formula** (of type t).

- Whenever P is of type $\langle e, t \rangle$ and t is of type e then $P(t)$ is a **formula** (of type t).

- Standard propositional logic.

- Whenever ϕ is a formula, v an e-variable and Q an e-quantifier then $Qv\phi$ is a **formula**. In particular $Some\,v\phi$ is **true** if ϕ is true for some v in E and $Every\,v\phi$ is **true** if ϕ is true for each v in E.

- Whenever ϕ is a formula and v an e-variable (o-variable) then $\hat{v}[\phi]$ is an e-set term (o-set term). It denotes the set of elements in E (O) that satisfy ϕ.

So far the system is little more than a notational variant of first order logic with respect to the domain E. The larger domain O is not exploited yet. We have chosen mnemonic names for the quantifiers *some* and *every*. This is done to indicate that they will represent the corresponding words in English, say in *some girl* and *every boy*. These quantifiers are here introduced as unary but could as well have been introduced as binary to correspond better to the linguistic form (Barwise and Cooper 1981). The definition used here extends in a natural way to generalized quantifiers (see Lønning 1987b, 1989a).

To represent plural NP's, the following additional symbols and rules are included.

- The language includes an operator $\oplus$ of type $\langle o^2, o \rangle$: If t and s are terms then so is $t \oplus s$; $\oplus$ denotes a function from $O \times O$ into O accordingly.

- It contains an operator τ of type $\langle \langle e, t \rangle, o \rangle$: When S is an e-set term then $\tau(S)$ is a **term**; τ denotes a function from the power set of E into O.

- It contains an operator $*$ of type $\langle \langle e, t \rangle, \langle o, t \rangle \rangle$: When S is an e-set term then $^*(S)$ is a **set term**; $*$ denotes a function from the power set of E into the power set of O.

- It contains an operator D of type $\langle \langle o, t \rangle, \langle o, t \rangle \rangle$: If S is a set term then $^D(S)$ is a **set term**; D denotes a function from the power set of O into the power set of O.

- It contains a quantifier $\exists$. When ϕ is a formula and v an o-variable then $\exists v\phi$ is a **formula**. The formula is **true** if ϕ is true for some v in O.

To see how these new symbols work, we shall show how some of the original example sentences now can be represented.

(6) (a) *Some $z[boat(z) \wedge bought(john \oplus harry, z)]$*

    (b) *Some $z[boat(z) \wedge bought(\tau(boy), z)]$*

    (c) *Some $x[boy(x) \wedge Some\ y[girl(y) \wedge Some\ z[boat(z) \wedge bought(x \oplus y, z)]]]$*

    (d) *$\exists x[Some\ z[boat(z) \wedge {}^*boy(x) \wedge bought(x, z)]]$*

    (e) *$\exists x[Some\ z[boat(z) \wedge {}^*boy(x) \wedge bought(john \oplus x, z)]]$*

The symbol $\oplus$ represents the collective conjunction. The noun phrase *John and Harry* in sentence (1a) is represented by $john \oplus harry$. This refers to the object that consists of John and Harry and will in the model denote some element in O. The symbol τ represents the plural definite article, so *the boys* is represented by $\tau(boy)$. Semantically, the definite article turns the set of boys into the object that consists of all boys. The symbol $\oplus$ is also sufficient for representing the conjunction of indefinite singular noun phrases, like (6c).

To interpret collectively read plural indefinite NP's, like *some boys, three girls*, ∃ is introduced as the existential quantifier over the set O. Observe that *Some* is only used in PL to model singular noun phrases. The plural *some girls* is represented by the use of ∃, at least when it is read collectively. The existential quantifier is not thought of as part of the indefinite NP, but it is introduced by a larger part of the sentence or discourse. What the indefinite plural noun phrase contributes, e.g. *some girls*, is a restriction on the variable bound by that quantifier, here that it is a collection of girls. This is represented by *girl*. Similarly, *three girls* restrict the bound variable both to be a collection of girls and to have a specified size. While *girl* in PL denotes a set of individuals, *girl* denotes the set of collections that have these individuals as constituents. The star operator is not meant to correspond directly to syntactic plural formation. It will not be used in the representation of plural NP's that must be read distributively, like *most girls*, nor will it be used in the representation of definite NP's. We will not in this paper discuss how the two semantic forms S and *S should be related to syntactic plural. Confer Frey and Kamp (1990) for a discussion of problems related to this.

The D-operator is introduced to tackle examples like (7a). Here the subject NP gets a collective reading with respect to the first VP and a distributive reading with respect to the second VP. To interpret this with a uniform semantic interpretation of the subject NP we must choose a collective interpretation of the NP. The D-operator will then recover the individual parts from the collection.

(7) (a) The boys carried the piano upstairs (together) and got a cookie (each).

(b) $\hat{z}[\hat{x}[Some\ y[piano(y) \wedge carried\ upstairs(x,y)]](z) \wedge$
$^D(\hat{x}[Some\ y[cookie(y) \wedge got(x,y)]])(z)](\tau boy)$

The role of the D-operator is to take a property, here $\hat{x}[Some\ y[cookie(y) \wedge got(x,y)]]$ and turn it into a new property $^D(\hat{x}Some\ y[cookie(y) \wedge got(x,y)])$ which is true of a collection if the original property is true of each of the individuals that constitute the collection, i.e. in the example, true of each boy.

Historical remarks. The notation used here is chosen close to the one used by Link (1983), although which symbols are primitive and which ones defined depart from that paper. The idea to consider the operators as basic functions without specifying their semantics further can be traced back to Blau (1981), but he only considered the definite article, here represented by τ. To consider collections as object of the same logical type as ordinary individuals was prominent in the Lesniewskian mereology and Leonard and Goodman's (1940) "calculus of individuals" which was applied in a more linguistic setting by Massey (1976). Also the way Scha (1981) uses sets to model plurals falls within this common framework.

Conditions on the Semantics

The semantics for the new operators introduced so far is the most general one. It subsumes all the proposals from the literature. It is thereby also the logically weakest one: it makes few inferences valid. In particular, the semantics does not relate the denotations of the two noun phrases *John and Harry* and *Harry and John*. Conversely, models are allowed where *John, Harry, Dick* and *Bill* denote four different individuals and *John and Harry* and *Dick and Bill* denote the same object.

We will consider more limited classes of model structures — or, in other words, we will consider additional conditions on what counts as valid interpretations of the new symbols — which will better correspond to our intuitions of what tne right semantics of plurals is. We will here rely on what we can call "ontological intuitions". We have a pretty clear picture that whatever the object referred to by *John and Harry* is, it is the same as the object referred to by *Harry and John*. Immediately, this seems to be grounded in the fact that the objects referred to by the two phrases have the same physical extension. But this is not sufficient. *John* and *John's body* are physically co-extensional, still on a logical semantic analysis we will ascribe different denotations to these two phrases. The reason is that there are properties that the one has that the other does not have. Similarly, when we turn to plural NP's. Two NP's shall be ascribed the same denotation only if they can be ascribed exactly the same properties (on a collective reading), or, in other words, if they are mutually substitutable.

We will consider some specific examples of restrictions on the semantics of plural NP's. For a more complete survey see Lønning (1989a). We start with the interpretation of $\oplus$. We will write $+$ for the denotation $[\oplus]$ of $\oplus$ in the sequel. There are three restrictions that might be imposed.

Commutativity: for all a and b in O: $a + b = b + a$

Idempotency: for all a in O: $a + a = a$

Associativity: for all a,b, and c in O: $(a + (b + c)) = ((a + b) + c)$

Commutativity corresponds to the basic ontological intuition that whatever the status is of the object to which *John and Harry* refers, this is the same object as the one to which *Harry and John* refers. Possible counter-examples are *John and Mary are husband and wife* or the situation where the authors of a scientific paper are listed in a non-alphabetic order. We think the conjunction here is used in different ways than in the collective readings. The NP does not refer directly to a collection. The more complex interpretation is determined from convention (the "paper" example) or it is a *respectively* construction (the "husband and wife" example) which will have to be resolved at a more syntactic level.

It is harder to find examples to motivate idempotency simply because we do not say things like *John and John* and refer to the same person both times we utter *John*. But PL generates such terms. Of course we could exclude terms like this syntactically, but we do not want to exclude the possibility that two different names, *John* and *Harry*, refer to the same individual. The intuition behind the constraint then is that the noun phrase *John and Harry* in this situation does not have any other reference than *John*. This becomes even clearer in a situation where, say, *the lawyers* and *the doctors* are co-referential. Then these again refer to the same object as *the doctors and the lawyers*.

As for associativity, the following two NP's seem to refer to the same object and be mutually substitutable.

(8) (a) John, Harry and Bill

    (b) Bill, John and Harry

It seems reasonable to think of *a, b and c* as shorthand for *a and b and c*, as we do for other phrases, like sentences and VP's. Whether the NP in (8a) then shall be represented as $john \oplus (harry \oplus bill)$ or $(john \oplus harry) \oplus bill$ is without importance if + is **associative**. This, together with the commutativity, will ascribe the same denotation to the two NP's in example (8), and in general have the effect that the order is without importance in collective NP's of the form *a,b,c,..., and z*.

    This constraint has been much more disputed than the other two, however. It has been argued that there are sentences where the grouping has an impact (Hoeksema 1983):

(9) (a) (Blücher and Wellington) and Napoleon fought against each other near Waterloo.

    (b) Blücher and (Wellington and Napoleon) fought against each other near Waterloo.

There is a difference between Wellington fighting with Blücher against Napoleon or with Napoleon against Blücher.

    There are two different approaches to the interpretation of the collective *and*. The first one is to say that the collective conjunction has the three properties mentioned. One tradition in the interpretation of collectively read plural NP's has been to use domains which are semi-lattices (or special types of such semi-lattices, e.g. complete atomic Boolean algebras) and where *and* is interpreted as the lattice join-operation (Link 1983). It is a standard observation from lattice theory that a binary operator has the three properties: commutativity, idempotency, associativity if and only if it is a join operator in a semi-lattice. Hence, *and* should be interpreted as a join operator if and only if the three properties are reasonable. Link's (1983) use of complete atomic Boolean algebras, the "calculus of individuals" (Leonard and Goodman 1940, Scha 1976) and Scha's (1981) use of sets where *John* denotes a singleton $\{j\}$ and *John and Harry* denotes $\{j, h\}$ are all examples of more restricted classes of semi-lattices. Observe also that all other interpretations of *and* in English, e.g., sentence and VP conjunction, have the three properties. For a discussion of how sentences like (9) can be interpreted in models where the collective conjunction has the three properties, we must again refer to Lønning (1989a).

    The other possibility has been to say that the conjunction is commutative, idempotent (or undefined when the two arguments are equal), but not associative (Hoeksema 1983, Landman 1988, 1989) or ambiguous between an associative and a non-associative reading (Link 1984). We will not go into the details of these proposals here (for a comparison of them, see Lønning 1989a), but we will make some general observations. The first point is that a conjunction of more than two arguments does not always group the arguments in pairs. In particular no grouping is possible for *Blücher, Wellington, and Napoleon*. It is hence natural to extend the language PL with *n*-ary operators $\oplus_n$ for $n > 2$ and represent this last NP by $\oplus_3(b, w, n)$. With the explicit repetition of *and*, the NP *Blücher and Wellington and Napoleon* is three times ambiguous: between

$\oplus_3(b, w, n)$ and the two "grouped" readings $\oplus_2(\oplus_2(b, w), n)$ and $\oplus_2(b, \oplus_2(w, n))$. Similarly, an NP of four conjuncts, like *Tom and Dick and Harry and Bill*, will be 11 times ambiguous. In speech one will normally use prosody to disambiguate between all these options.

Even though the mentioned proposals do not assume the conjunction to be associative, the formalisms used for interpreting the conjunction, e.g. sets in different fashions, restrain the behaviour of the conjunction beyond commutativity and idempotency. One condition common to all the proposals — also the associative ones — is the following. In the model structure there is a function F with domain all finite subsets of O and range elements of O such that for all $n \geq 2$ and $t_1, \ldots, t_n$, $[\oplus_n(t_1, \ldots, t_n)] = F(\{[t_1], \ldots, [t_n]\})$. So the collective conjunction is not more specific than set formation; it does not take into consideration the order of its arguments nor repetitions of arguments.

We have so far treated only $\oplus$. Which additional restrictions should be put on the interpretation of the symbols? If John, Bill and Harry are boys and the only boys in a situation then the term *the boys* intuitively refers to the same as the term *John, Bill and Harry* and the two sentences (10) express the same, i.e. entail each other.

(10) (a) John, Bill and Harry bought a boat together.

    (b) The boys bought a boat together.

If we write T for the function that interprets τ and interpret conjunction in terms of a function F as explained in the last paragraph above, then T and F should agree on their common domain: finite non-empty subsets of E. This assumption is shared by all authors who consider both collective conjunction and plural definite descriptions.

This does not entail anything for the behaviour of F when one of the arguments belong to $O \setminus E$, nor does it entail anything for T when the argument is an infinite subset of E. If the conjunction is constrained to have the three properties we have discussed, $T(X)$ will be the semi-lattice join of X whenever X is a finite subset of E. It is then reasonable to also let $T(X)$ be the join or supremum of X when X is an infinite subset of E. For this purpose the underlying lattice structure has been constrained to be a complete semi-lattice (Link 1983).

There is also a natural connection between definite and indefinite NP's. The sentences in (10) intuitively entail the two sentences:

(11) (a) Some boys bought a boat together.

    (b) Three boys bought a boat together.

This means that $\tau\hat{x}[\phi]$ should be a member of $[*boy]$ if $\hat{x}[\phi]$ is a subset of $[boy]$. Conversely, only elements in O of the type $T(X)$ for X a subset of $[boy]$ should be members of $[*boy]$. This yields a connection between the interpretation of the symbols τ and $*$. We will return to some of the details of this connection later on.

Free Structures

The additional conditions considered so far relate the interpretations of the different symbols and secure that certain NP's get the same denotations as other NP's. So far we have not considered the dual question: how to prevent *John and Harry* from denoting the same as *Dick and Bill*. For simplicity, we concentrate here on the definite descriptions and drop the collective conjunction. The strongest possible additional property we can put on the interpretations is that the function T is injective; that $T(X) = T(Y)$ only if $X = Y$. (We here assume that the domain of T is the set of subsets of E of cardinality at least two. For a discussion of singleton and empty sets, see Lønning, 1989a).

An interesting observation is that if T is a supremum operator on a semi-lattice then it is injective if and only if O is a complete atomic Boolean algebra with E the set of atoms. Link (1983) proposed to use such structures and Scha's (1981) use of sets are isomorphic to this type of structures. These kinds of model structures are hence appropriate if and only if one accepts the associativity of the conjunction and the injectivity of the definite descriptions.

Is it appropriate to restrict T to be injective? The opinions differ. In the nominalist tradition the reference of *the boys* was identified with their physical extension either as a heap of atoms or as a part of space-time. *The windows* and *the windowparts* have the same physical extension, hence T applied to the two different sets [*window*] and [*windowpart*] should yield the same value (Leonard and Goodman 1940). But in a more linguistic logical setting, it is not difficult to find properties we ascribe to one of the two objects and not to the other one; just like it is not difficult to find properties that *John* has and *John's body* does not have: *to be a human, to be married, to own a car* .... From a logical point of view, *the windows* and *the windowparts* are different objects, and this has been the most common view within proposed semantics for plurals.

But there might be other examples indicating that the assumption about injectivity is too strong. The so-called "group-denoting" nouns, like *pair, couple, set, group*, have several syntactic and semantic particular properties. They can in the singular number agree with verbs in the plural in British English, *the committee are...*; they can occur in contexts like *a ... of men*. Semantically, the two sentences

(12) (a) The group carried the books upstairs.
　　 (b) The members of the group carried the books upstairs,

seem to bring exactly the same information. One might want to identify *the group* and *the members of the group*. In other contexts one might similarly want to identify *the groups* and *the members of the groups*. Hence one will allow T to get the same value for the two different sets [*group*] and [*member of the groups*] (remember that [*the groups*] $= T([group])$).

It turns out that how one wants to treat this question is a real branching point between different proposed semantics for plurals. We have already mentioned several proposals that let T be injective. Conversely, in a later proposal Link (1987) used more general lattices which allow an identity between a group and its members, and one main point in Landman's (1989) proposal is also to allow identity.

There is one point where an identification between *the groups* and *the members of the group* has undesirable consequences. We have so far not discussed which additional restrictions to put on the interpretation of the distributivity operator D. But it is clear that we want it to work such that the interpretation of (13a) as (13b) is equivalent to (13c):

(13) (a) The members of the groups got an apple (each).

    (b) $^D\hat{x}[Some\ y[apple\,(y) \land get\,(x,y)]](\tau(member\_of\_the\_groups\,))$

    (c) Each member of the groups got an apple.

To obtain this, it must be possible to recover the individual group members from the object *the members of the group*; T must be injective. If not, if *the members of the groups* equals *the groups*, sentence (13a) and (14a) will get the same interpretation, but (13c) and (14b) are not equivalent.

(14) (a) The groups got an apple (each).

    (b) Each group got an apple.

There seems to be two conflicting objectives when we come to group-denoting nouns. In some contexts we want to identify a group with the collection of its members (12); in other contexts we want to be able to separate the two (13), (14). So far no proposal for a formal semantics of plurals has been able to meet both considerations. It seems that the way out is to be less dogmatic about the relationship between linguistic form and semantic interpretation and admit that some world knowledge enters the picture when humans use natural languages.

Questions of Logic

How can these observations about the semantics of plural NP's be put to work in text processing? First, we have to try to find proof procedures corresponding to the concept of logical validity. It has been argued that the semantics of plural NP's shows that the underlying logic of English is that of second order logic and hence that the semantics of English is not computable. Does this also apply to the logic PL?

The first observation is that PL with respect to the most general semantics we sketched at the beginning is an innocent extension of first order logic: it is complete and compact. The proof is based on the following observations (for details of the proof, see Lønning 1989a):

- If first order logic is extended with some second order predicates $\mathcal{P}_1, \ldots, \mathcal{P}_n, \ldots$ and the formation rules:

    - If ϕ is a formula and v a variable then $\hat{v}[\phi]$ is a set term (unary predicate).
    - If S is a set term and t is a term then $S(t)$ is a formula.
    - If $\mathcal{P}$ is a second order predicate and S a set term then $\mathcal{P}(S)$ is a formula,

    then an axiomatization of first order logic extended with

    - For all set terms $\hat{v}[\phi]$ and terms t free for v in ϕ, $\hat{v}[\phi](t) \leftrightarrow \phi(t/v)$.
    - For all set terms S and T and second order predicates $\mathcal{P}$,
      $\forall x(S(x) \leftrightarrow T(x)) \rightarrow (\mathcal{P}(S) \rightarrow \mathcal{P}(T))$,

    yields completeness and compactness. (This is shown implicitly by Keisler 1970).

- The result carries over to languages with relation symbols with arity (m, n), i.e. which take m terms and n set terms as arguments.

- Function symbols that take m terms and n set terms as arguments and give terms as values can be eliminated in terms of relation symbols with arity $(m + 1, n)$. Similarly, functions that give set terms as values can be eliminated.

- We see that PL is exactly such an extension of first order logic. In addition, the domain is sorted, but in our context this does not introduce any new problems.

Most of the additional conditions we have considered can be expressed with axioms or axiom schemata within PL. Thus compactness and completeness are retained. For example, the fact that $\oplus$ is commutative is simply expressed by:

(15) For all terms t, s: $t \oplus s = s \oplus t$

There is one type of restriction that cannot be expressed, however. If $[^\bullet \hat{x}[\phi]] = \{T(X) : X \subseteq [\hat{x}[\phi]] \ \& \ X \neq \emptyset\}$ and T is total and injective then $[^\bullet \hat{x}[\phi]]$ will be in a one-to-one correspondence with the power set of $[\hat{x}[\phi]] \setminus \emptyset$. It follows that the whole domain O will correspond to the power set of E and that the quantifier $\exists$ will correspond to second order quantification in a sense made precise in Lønning (1989a). Logically, with this restrictions on the semantics, the logic PL becomes very similar to an interpretation of plurals through second order logic, and the logic gets the same power as monadic second order logic. Also, if T is not injective, the logic gets this power if $T(X)$ is the lattice supremum of X for all subsets X of E.

We can make a small modification in the semantics, however. Call a subset X of O **definable** provided it equals a denotation $[\hat{v}[\phi]]_g$ for some set term $\hat{v}[\phi]$ and variable assignment g. Call a structure **definable** provided T is not a total but some partial function from the power set of E into O, which has at least all the **definable** subsets of E in its domain. T shall be the interpretation of τ as before and the interpretation of $*$ will be $[^\bullet S] = \{y \in O : \exists X \subseteq E(T(X) = y)\}$. With this modification it is possible to maintain a complete axiomatization even if T is injective or the lattice supremum operator. (This corresponds to a generalized completeness for the second order logic).

The most correct semantics for plural NP's seem to be the one we get when T is total. In principle, any set of boys should be permitted to form a collection that will be the witness for *some boys*. On the other hand, if T is permitted to be partial as explained in the paragraph above, the result is a much simpler logic. Are there English sentences that get affected if we use definable structures instead of total structures? Call the sentences that are not affected **persistent**. Observe that the definite plural NP's are persistent; τ gets exactly the same type of interpretation in the two classes. It is only sentences that contain $*$ that might not be persistent. One possible example of a sentence that is not persistent is the following.

(16) (a) There are some horses that are all faster than Zev and also faster than the sire of any horse that is slower than all of them.

(b) $\exists X(\exists z(X(z)) \wedge \forall z(X(z) \rightarrow f(z,0)) \wedge$
$\qquad \forall y(\forall z(X(z) \rightarrow f(z,y)) \rightarrow \forall z(X(z) \rightarrow f(z,s(y)))))$

(c) $\exists x[^{\bullet}\hat{y}[y=y](x) \wedge {}^{D}\hat{y}[f(y,0)](x) \wedge$
$\qquad {}^{D}\hat{z}[Every\, y[{}^{D}\hat{u}[f(u,y)](x) \rightarrow f(z,s(y))]](x)]$

The example sentence (16a) is from Boolos (1984) who translates it into second order logic as (16b) presupposing quantification over horses and using 0, s, f for *Zev, the sire of, is faster than*, respectively. Formula (16c) is a similar representation in PL. The formulas in the example are not persistent. As Boolos points out, if (16b) is reinterpreted in standard arithmetic, reading s as the successor function and f as *greater than*, it becomes false in the standard model of arithmetic and true in all non-standard models.

The first reaction to the sentence is that it is odd. We must make an effort to understand it in the intended way. The sentence is more acceptable from the speaker's point of view, however. We can imagine a speaker using the sentence if he has a specific collection in mind, to which he wants to refer. And he refers to this collection both by the use of *some horses* and by the use of *them*. In particular, we imagine that Boolos (1984) when he constructed the sentence, had numbers rather than horses in mind and made a reference to the non-standard numbers (or to any non-wellfounded set of such numbers).

There has been a long discussion of the right interpretation of definite descriptions. In particular, Donellan (1966) has challenged the long-standing Russelian view and claimed that definite descriptions can be used in two ways: The attributive (Russelian) use and the referential use, in which the speaker makes a genuine reference to one particular object. We will not enter the discussion of the finer details of this view, but observe that it has been proposed to extend it also to indefinite NP's (Fodor and Sag 1982). It is not obvious how the referential readings shall be represented in a logical calculus like first or second order logic or PL. Suppose that the speaker by an utterance of the sentence *A friend of mine has talked to the president* with the description *a friend of mine* wants to refer to Jones. Suppose furthermore that Jones has not talked to the president but that another friend of the speaker, Smith, has talked to the president. Is the utterance true or false? If our understanding of the referential use makes the utterance false then the closest we come to a logical representation is to represent *a man* by the name *Jones*. A similar approach to (16a) would be to introduce a new name for the object referred to. Instead of the second order formula (16b) we would introduce a new unary predicate, say P, and represent it

(17) $\exists z(P(z)) \wedge \forall z(P(z) \rightarrow f(z,0)) \wedge$
$\qquad \forall y(\forall z(P(z) \rightarrow f(z,y)) \rightarrow \forall z(P(z) \rightarrow f(z,s(y))))$

This formula is first order, however. Similarly, if the variable x in (16c) is exchanged with a name, the result is a persistent formula. In the plural case, the distinction between attributive and referential readings and how the referential readings are conceived have dramatic consequences for the logical complexity of the logical formulas that are used for representing English sentences. This is in marked contrast to the standard singular/first order case.

There have been several other types of example sentences involving collectively read NP's that are claimed not to be persistent. We will not here discuss them in detail. But it seems that whether or not they are persistent depends on how we conceive other semantic components, like the referential reading exemplified above (see Lønning 1989a for a discussion). Regardless of whether English contains sentences that are not persistent, we have a complete axiomatization for the subset of English that consists of only persistent sentences. The axiomatization with respect to the definable structures will also be complete with respect to the structures where T is total. The axiomatization will be an extension of first order logic. It is an open question how effective proof procedures and procedures for automatic deduction can be extended to these systems.

We have only taken a few steps toward "intelligent" text processing!

References

Barwise, J.: 1989, *The Situation in Logic*, CSLI Lecture Notes No. 17.

Barwise, J. and R. Cooper: 1981, 'Generalized Quantifiers and Natural Language', *Linguistics and Philosophy* **4**, 159–219.

Barwise, J. and J. Perry: 1983, *Situations and Attitudes*, Bradford Books, Cambridge.

Bennett, M.R.: 1975 'Some Extensions of a Montague Fragment of English', thesis, distributed by the Indiana University Linguistics Club.

Blau, U.: 1981, 'Collective Objects', *Theoretical Linguistics* **8**, 101–130.

Boolos, G.: 1984, 'To Be Is to Be a Value of a Variable (or to Be Some Values of Some Variables)', *Journal of Philosophy* **81**, 430–449.

Bresnan, J. (ed.): 1982, *The Mental Representation of Grammatical Relations*, MIT Press.

Cohn, A. G.: 1987, 'A More Expressive Formulation of Many Sorted Logic', *J. Automated Reasoning* **3**, 113-200.

Colban, E.: 1987, 'Prepositional Phrases in Situation Schemata', published as Appendix A to Fenstad et al. (1987).

Donellan, K.S.: 1966, 'Reference and Definite Descriptions', *Philosophical Review* **75**, 281–304.

Dyvik, H.: 1988, 'Sentence Synthesis from Situation Schemata: A Unification-based Algorithm', *Nordic Journal of Linguistics* **11**(1-2), 17-32.

Fenstad, J. E.: 1988, 'Natural Language Systems', in *Advanced Topics in Artificial Intelligence*, R. T. Nossum (ed.), Springer-Verlag.

Fenstad, J. E., P. K. Halvorsen, T. Langholm, and J. van Benthem: 1985, 'Equations, Schemata and Situations', CSLI Report No. CSLI-85-29. Preliminary version of Fenstad et al. (1987).

Fenstad, J. E., P. K. Halvorsen, T. Langholm, and J. van Benthem: 1987, *Situations, Language and Logic*, Reidel, Dordrecht.

Fenstad, J. E., T. Langholm, E. Vestre: 1990, 'Representations and Interpretations', in *Proceedings of the Tutorials and Workshop on Formal Semantics and Computational Linguistics. Lugano, August 1988.*

Fodor, J.D. and Sag, I.A.: 1982, 'Referential and Quantificational Indefinites', *Linguistics and Philosophy* 5, 355–398.

Frey, W. and H. Kamp: 1990, 'Some Problems about Definite Plural Noun Phrases', in this volume.

Gärdenfors, P.: 1987, 'Induction, Conceptual Spaces and AI', in *Proceedings of the Workshop on Inductive Reasoning*, Risø National Lab., Roskilde, Denmark.

Halvorsen, P. K.: 1988, 'Situation Semantics and Semantic Interpretation in Constraint-Based Grammars', in *Proceedings of the International Conference on Fifth Generation Computer systems, FGCS-88, Tokyo, Japan, November 1988.*

Halvorsen, P. K. and R. M. Kaplan: 1988, 'Projections and Semantic Description in Lexical-Functional Grammar', in *Proceedings of the International Conference on Fifth Generation Computer systems, FGCS-88, Tokyo, Japan, November 1988.*

Hoeksema, J.: 1983, 'Plurality and Conjunction' in A.G.B. ter Meulen (ed.), *Studies in Modeltheoretic Semantics*, Foris, Dordrecht, 63–83.

Johnson, M.: 1988, *Attribute-Value Logic and the Theory of Grammar*, CSLI Lecture Notes No. 16.

Johnson, R., M. Rosner and C. J. Rupp: 1990, 'Situation Schemata and Linguistic Representation' in *Proceedings of the Tutorials and Workshop on Formal Semantics and Computational Linguistics. Lugano, August 1988.*

Keisler, H.J.: 1970, 'Logic with the quantifier 'there exists uncountably many' ', *Annals of Math. Logic* 1, 1–93.

Landman, F.: 1988, 'Groups, Plural Individuals and Intentionality' in J. Gronendijk, M. Stokhof and F. Veltman (eds.), *Proceedings of the Sixth Amsterdam Colloquium. April 13-16 1987*, ITLI, University of Amsterdam, 197–217.

Landman, F.: 1989, 'Groups 1', *Linguistics and Philosophy* 12, 559–605.

Langholm, T.: 1989, 'Algorithms for Partial Logics', Cosmos preprint No 12, Dept. of Mathematics, University of Oslo.

Leonard, H.S. and N. Goodman: 1940, 'The Calculus of Individuals and its Uses', *Journal of Symbolic Logic* 5, 45–55.

Link, G.: 1983, 'The Logical Analysis of Plurals and Mass Terms: A lattice-Theoretical Approach' in R. Bäuerle, C. Schwarze and A. von Stechow (eds.), *Meaning, Use and Interpretation of Language*, Walter de Gruyter, Berlin.

Link, G.: 1984, 'Hydras: On the Logic of Relative Constructions with Multiple Heads' in F. Landmann and F. Veltmann (eds.), *Varieties of Formal Semantics*, Foris, Dordrecht, 245–257.

Link, G.: 1987, 'Generalized Quantifiers and Plurals' in P. Gärdenfors (ed.), *Generalized Quantifiers. Linguistic and Logical Approaches*, Reidel, Dordrecht, 151–180.

Lønning, J.T.: 1987a, 'Mass Terms and Quantification', *Linguistics and Philosophy* 10, 1–52.

Lønning, J.T.: 1987b, 'Collective Readings of Definite and Indefinite Noun Phrases' in P. Gärdenfors (ed.), *Generalized Quantifiers. Linguistic and Logical Approaches*, Reidel, Dordrecht, 203–235.

Lønning, J.T.: 1989a, 'Some Aspects of the Logic of Plural Noun Phrases', thesis, Dept. of Mathematics, University of Oslo, Cosmos preprint no 11.

Lønning, J.T.: 1989b, 'Computational Semantics of Mass Terms', in *Proceedings of the Fourth Conference of the European Chapter of the Association for Computational Linguistics*, Manchester, April 1989, 205–211.

Massey, G.: 1976, 'Tom, Dick and Harry, and all the King's men', *American Philosophical Quarterly* 13, 89–107.

Meier, J, D. Metzing, T. Polzin, P. Ruhrberg, H. Rutz and M. Vollmer: 1988, 'Generierung von Wegbeschreibungen', KOLIBRI Arbeitsbericht nr. 9, DFG-Forschergruppe Kohärenz, Fakultät für Linguistik und Literaturwissenschaft der Universität Bielefeld. In German.

Pinkal, M.: 1990, 'On the Logical Structure of Comparatives', this volume.

Pollard, C. and I. Sag: 1987, *Information-Based Syntax and Semantics*, CSLI Lecture Notes No. 13.

Scha, R.: 1981, 'Distributive, Collective and Cumulative Quantification' in J.A.G. Gronendijk, T.M.V. Janssen and M.B.J. Stokhof (eds.), *Formal Methods in the Study of Language*, Mathematical Centre Tracts 136, Amsterdam, 483–512.

Sells, P.: 1985, *Lectures on Contemporary Syntactic Theories*, CSLI Lecture Notes No. 3.

van Benthem, J.: 1982, *The Logic of Time*, Reidel, Dordrecht.

Propositional and depictorial representations of spatial knowledge: The case of *path*-concepts#

Christopher Habel
University of Hamburg
Computer Science Department

Knowledge representations are of central importance in natural language processing systems, and spatial representations have a central position in knowledge representation. The reasons for this are described, and then a brief example is given to justify the assumption of a a dual-coding approach for the processing of spatial expressions. This means that beyond propositional representations a second, non-propositional representation format is used, that of depictorial representations. These representations are studied in the present paper with respect to some of their topological properties.

In Artificial Intelligence, the topological concept of *path* is usually seen from the viewpoint of trajectories; thus verbs of motions are normally considered with respect to the path concept. In section 2 I show that paths are a powerful concept for handling prepositions, too; the case of between is studied. This preposition is chosen, because it is usually not combined with verbs of motion. In other words, the usefullness of paths is demonstrated in static constellations, too.

In the final section I sketch the application of the results from the PP-study on the verb of motion follow.

1. Preliminaries on the Representation of Spatial Knowledge

Knowledge representations are of central importance in natural language processing systems because they are the goal and the foundation of communicative processes and, moreover, because they are the basis of reasoning processes, cf. fig. 1. This means that knowledge representations are the building blocks from which internal models of the external world are constructed.[1]

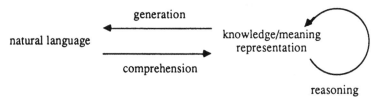

Fig. 1: The centrality of knowledge representations

This research has been carried out in close connection with the LILOG-project, which is funded by IBM-Germany. I thank Carola Eschenbach, Michael Herweg, Siegfried Kanngießer, Simone Pribbenow, Klaus Rehkämper and Geoff Simmons (who also corrected my English) for stimulating discussions and for their comments on an earlier version of this paper.

[1] The external world is not assumed to be "the real world". Fictional worlds are also a source for building internal, i.e. mental, models. Since the status of internal models is not the main topic of the present paper, I merely refer to Johnson-Laird (1983) and Habel (1986).

The system of knowledge representation determines what can be "spoken or thought" about. This means: the capacity of representation formalisms, for instance the storing, manipulation and acquisition of the entities and structures of the representation language, determines the efficiency of a knowledge-based system. The main topic of the present paper involves a special field of knowledge, namely the way that spatial knowledge can be represented and processed adequately, and within this specific type of knowledge one concept, namely that of a *path*, is investigated in detail.

There are several reasons for an focussing on this particular field of spatial knowledge: one of them is its "ubiquity". This can be appreciated best by considering where spatial knowledge is of importance. The answer is: "(almost) everywhere, because human behavior (and therefore communication, planning etc. as well) is grounded in space and time."

Cognitive and computational linguistics face the problem of cognitively adequate representation of spatial information in the same way as Artificial Intelligence. Since the description and explanation of linguistic processes in human beings is the primary aim of cognitive linguistics, the representation of text meanings and of background knowledge (world knowledge) used in text understanding and text generation has to be an essential topic of research. The same holds for computational linguistics and language-oriented AI, even if the reason for focussing on these topics is sometimes motivated by the needs of application.

"Representation of spatial knowledge" as a subject of study is also of particular relevance because here we find the "intersection" of two relevant fields of information processing abilities, namely of language processing and of image processing, cf. fig. 2.

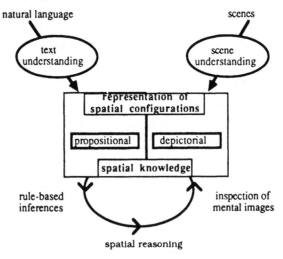

Fig. 2: The central position of spatial representations[2]

---

[2] Habel (1987) contains this figure and an extended argumentation for a dual-coding theory (see below).

An information processing system equipped with both language and image understanding abilities will above all be organized in such a way that analogue information used in the process of image understanding can be related to the propositional representations emerging from language processing in a well-defined way. This means - among other things - that we must determine whether non-propositional representations, which I call - following Kosslyn (1980) - *depictorial* (see below) are produced from linguistic inputs as well. Before I discuss the properties of such non-propositional representations in detail, I will give a sketchy argument for depictorial representations in language processing. The brief example (1) concerning spatial prepositional phrases gives evidence for the usefulness of depictorial representations.

(1) a. Früher standen zahlreiche alte Weiden um den Teich im Stadtpark. (Long ago there were many old willows around the pond in the city park.)

    b. In diesem Herbst wurden die Weiden auf der West- und der Südseite gefällt.(This autumn the willows on the west and on the south side were felled.)

    c. Jetzt stehen noch alte Weiden am (Ost- und Nordrand des)Teich(es) im Stadtpark. (Today there are still old willows on (the east and north side of) the city park pond.)

For the sentence (1.a) a *mental image* ,i.e. an internal representation - depicted here by a sketch as shown in fig. 3.a - may be assumed.[3]

The core of the prepositional concept <u>um</u> (<u>around</u>) is characterized by the fact that the objects denoted in the <u>um</u> -NP - in our example the old willows - induce a path-like object with the topological property of being *closed*, i.e. a *loop*, (see fig. 3.a). Since paths and loops are genuinely spatial concepts, a spatial representation format should be considered for the representation of <u>um</u>-NPs.

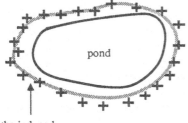

the induced
closed path

Fig. 3.a: The willows around the city park pond (with induced path)

Fig. 3.b: The pond in the city park after felling some trees

By the next sentence (1.b), a change to another mental image (like the one in fig. 3.b) is caused.(Note that the path induced by the willows is not closed any more.) The changed

---

3 Mental images and mental maps are explained more extensively below. Some problems of detailing such maps can be seen clearly in the example; certainly the depiction will not be so detailed that the exact form of the pond and the exact number of trees are "coded" in this knowledge entity (at least not usually, i.e. for an average speaker/hearer).

situation leads to a change in the natural language description, for example to (1.c). The preposition am (on), which does not require the loop-condition (closedness) of the induced path, may be used; um (around) cannot be used anymore, since the information of the second sentence (1.b) indicates that the path is no longer closed. This example demonstrates how the meaning of two spatial prepositions is distinguished by means of genuinely spatial concepts.

Non-propositional representations are - as I am convinced, cf. Habel (1987) - relevant for text understanding; furthermore the research field "representation and processing of spatial knowledge" is a field where the interaction of cognition and perception plays an important role. This configuration justifies - for cognition-oriented and computational linguists as well as researchers in Artificial Intelligence - intensive research on this subject matter.

As can be seen in fig. 2, we can imagine two clearly distinguished formats of representation for the processing of spatial knowledge: first, a format of propositional representation; second, a non-propositional format, that of depictorial representations. It is not controversial that analogue representations play an important role in the process of visual *perception*, at least in the intermediate levels of image processing. However, the interesting but more controversial question is whether image-like representations, i.e. genuine spatial (e.g. analogue) representations, should be assumed on the level of cognitive representation , too. The properties of depictorial representations remains to be characterized. (This problem is beyond the scope of the present paper.) As Rehkämper (1988) demonstrates, *analogue* and *image-like* are not equivalent concepts.

I will consider these problems from another point of view, namely that of *cognitive* processes. In the last decade, philosophers and cognitive psychologists showed a reawakened interest in the phenomena of *mental images*. The discussion, called the imagery debate[4], centers on the question whether mental images, which are also called *pictorial notions* or *mental pictures*, exist, i.e. if they are cognitively real or if they are epiphenomena, as Pylyshyn (1981) asserts. The fields of AI and cognitive psychology face a slightly modified question: "Are mental images an adequate means for representing spatial knowledge?" If we take a look at the situation of an information processing system as sketched in fig. 2, the question is whether only propositional representations should be assumed for natural language texts, or whether (in certain cases) text meaning (or parts of it) should be represented depictorially as well. This means that mental images are seen as one way to construct internal models. A similar perspective - with respect to images and mental models - is the topic of chapter 7 of Johnson-Laird (1983).

Before I discuss more detailed questions about the existence of mental images, I would like to summarize the two trends within the imagery controversy:

---

[4] Cp. N. Block (ed., 1981).

descriptionalist	depictorialists
existence of only one format of representation, the propositional (Pylyshyn 1981)	existence of several formats of representation (propositional and depictorial) (Kosslyn 1980)

Then the problems of representation are:

development of propositional representations for spatial knowledge	development of depictorial and propositional representations for spatial knowledge interaction of depictorial and propositional representations

As this comparison shows, joint research (by proponents of each of the attitudes) is possible and necessary in many fields, despite all the differences of the controversial views. Both views are based on propositional representations; differences are found particularly in the question if and in which cases non-propositional, i.e. depictorial, representations are used additionally.

If a combination of propositional and depictorial representations is assumed, then corresponding approaches - following Paivio - are regarded as *dual coding* theories (cf. Paivio 1983)[5]. In the following, I will follow the line of dual-coding theories; thus I presuppose a hybrid representation system based on propositional as well as depictorial representations. This line of thinking is justified, since the assumption of the existence of non-propositional, depictorial representations in cognitive processes or in mental models bridges the gap between cognition and perception.

At this state of the discussion, some remarks on image-like representations are appropriate. Following Kosslyn (1980, p.33), I assume that depictions occur in an internal spatial medium, which provides the material that mental images are formed from. This internal spatial medium is a theoretical entity just as propositional representation languages are theoretical entities. For example, it is possible to use cell-matrices[6] for a concrete realization of this abstract entity. Such matrices, which are rectangular arrays of discrete values, can only be approximations of real depictions, i.e. mental images in human mind.

In the present paper I will investigate some of the properties of the spatial medium in an abstract manner, without reference to a concrete realization or implementation. This

---

[5] It should be noted here that other specific formats of representation, e.g. for smell, taste etc. are conceivable. Such approaches, which could be called "multi-coding theories", always involve the interaction of cognitive processes and processes of the different types of perception. When and if these questions will be relevant for AI (in its application-oriented fields) and cognitive linguistics cannot - in my opinion - be decided with sufficient certainty at present.

[6] This type of representation is described in more detail in Habel (1987) and Khenkhar (1988). Our approach is strongly influenced by Kosslyn's (1980) CRT-metaphor and by Codd's (1968) cellular automata.

means that not specific representations but the principles of the representation format are the topic of discussion. This spatial representation format will be abbreviated by SPATREF in the following.

Any adequate realization of depictorial representations depends on those properties of SPATREF we are committed to. Therefore we have to clarify the following points:

(2.a)   What is the status of the SPATREF?

(2.b)   Which are the relevant (formal) properties
         (wrt[7] the status mentioned in (2.a)) of the SPATREF?

As I will discuss in more detail below, SPATREF is seen as the frame for the (cognitive) processes on spatial representations. Thus we have to determine the geometrical and topological properties of SPATREF, and this is the answer to (2.b). In other words, the subject under investigation is the *mental topology* or *mental geometry* on SPATREF.

Before I continue with mental topology, I want to explain the status of the considerations described in this paper. For this we make a short detour to *visual space*, i. e. "the space that appears to us visually" as Westheimer (1987) defines it. Some relevant characteristics of this psychological concept are summarized in the following:

> "...It is a manifold of three dimensions in which objects are located in an ordered fashion. What metrical properties it has do not make it a space of constant curvature,...
> An individual's concept of spatial relationships is not an exact replica of the "outside world"....
> By abstractions, and by logical deductions from moment-to-moment sensory input and from memory traces, a concept of spatial relationships is built up, transcending immediate visual space and containing fewer contradictions:..."
> (Westheimer. 1987, 796)

It follows that it is interesting and relevant to investigate, firstly, the (formal) properties of visual space and, secondly, the status of the "concept of spatial relationships" Westheimer refers to in the final sentence (cited above).

With respect to the first problem, the properties of visual space, I want only to mention the work of Blank (1958) and Roberts & Suppes (1967). Blank proposes an axiomatics of binocular vision on the basis of hyperbolic geometry, while Roberts & Suppes describe the geometry of visual perception by means of a tolerance relation (cp. Roberts, 1973). More important than the specific proposed axiomatizations, whose adequacy can be disputed, is the spirit of these investigations. Roberts & Suppes characterize their approach to visual space as follows:

> "The other approach to primitive visual space is to try to recover its geometry from certain observables. We are interested in studying what structures are

---

7 wrt is the abbreviation standing for "with respect to".

compatible with our primitive visual perceptions, what relations are meaningful in our primitive visual space, etc. ... one whole collection of observables are our judgements of comparative distance, alignment, or betweenness, parallelism, etc." (Roberts & Suppes, 1967, p.180).

The line of my investigation is congenial to this. The long-term goal is to recover the geometry and topology of the spatial medium of cognition, SPATREF; I will call a realization on SPATREF a *cognitive space* (CSP), analogous to visual space. This type of research is strongly related to those AI-activities which try to develop systems of *commonsense knowledge*. Some of the analyses of section 2 are similar to those of Kautz (1985), whose work is in the commonsense tradition. The main difference between his approach and the one presented here is that the *path* concept of section 2 is more strictly oriented to topology.

As evidence (*observables* in the sense of Roberts & Suppes, 1973), I will use linguistic examples; thus the way language describes space is seen as a mirror of the internal structure of cognitive space or the spatial medium SPATREF.

What is represented or depicted by a mental image? It is obvious that only limited situations or objects can be the subject of one mental image. In other words, the internal model of an information processing system contains a large set of image-like knowledge entities, each limited in size and resolution. This assumption is also made by Kosslyn (1980), who conducted experiments with respect to *size, angle* and - indirectly - *grain* of mental images.

In the following sections, it is assumed that individual mental images are instantiations (realizations) of the SPATREF-frame; thus they follow the system of principles for cognitive spaces, that each cognitive space is limited in size and grain. To refer to cognitive spaces (in plural), i.e. to assume a collection of such entities, can easily justified by using the analogy to maps.[8] Usually geographical objects are represented (encoded) in more than one map. Each of these maps has specific properties, e.g. scale and projection style, depending on which of the aspects are to be represented.

In the present paper I will deal with some topological concepts only, namely different types of paths. From these examples, I will argue that mental topology is at least as important as mental geometry. The relevant differences between topology and geometry with respect to cognitive space involve especially phenomena of deformation and measurement. Whereas topological concepts, such as path and connectivity, are invariant under (a specific class of) deformation, geometrical concepts, such as straightnes of a line, length or angle, can easily be changed by transformations. Furthermore, a strictly topological space does not contain any metric information. (On the other hand, a metrical space always induces a topology.) Thus we have at least three levels to be studied: the topological, the metrical and the geometrical. Especially in the comprehension of texts, topologi-

---

[8] Cp. the research on *mental maps*, especially Downs & Stea (1977).

cal concepts are used primarily, and geometrical concepts only secondarily. This is parallel to the fact that the mental images built up during comprehension are less detailed with respect to the geometry than images constructed via visual space from real scenes in the perceptual process.

2. Prepositional phrases and paths[9]

2.1 The frame of the analysis

In the following two subdomains of the ontological system are primarily important,

D ~ the domain of objects
LR ~ the domain of locations, i.e.the space of local regions.

Since I assume that locations, i.e. local regions, as entities of the internal model have to be realized on a spatial medium, the relation between LR and SPATREF is one of *realization on*. One further complication has to been mentioned: Since I presuppose the existence of a structure (or collection) of mental maps, it would be more appropriate to see LR as set of domains (in plural). From this would follow as a further extension that REG assigns to an object x a set of local regions, each of them part of a different domain of locations, i.e. of different mental maps.

Such a line of investigation, which is based on the assumption of structured relations in a system of cognitive spaces (or mental maps or mental images), lies beyond the scope of the present paper. It is obvious that further operations - especially some of the *embedding type* - are needed to describe such hierarchical systems of depictorial representations. (Cf. on this problems: McNamara 1986, Habel 1987).

The two domains D and LR are connected by a (time-dependent and context-sensitive) mapping

(3.a)   REG : D → LR

which assigns to an object x its characteristic region REG(x).

Furthermore, I assume a general relation of local inclusion

(3.b)   LOC : LR x LR   → BOOL,

which can be interpreted as follows

      LOC (x, y)   ~   the region x is locally included in the region y.

This system of concepts is similar to that of Bierwisch (1988, p.17ff.).[10] Based on these notations the general structure of the meaning of spatial prepositions is given by

---

[9] The same topic - from a more linguistic point of view- is dealt with in Habel (1989).

[10] Note that the terminology is different, especially with respect to LOC. Since Bierwisch sees local inclusion as set-theoretical inclusion, he does not use a specific relation like the LOC. His "loc" corresponds to the REG used here.

(4) $\lambda\,y\,\lambda\,x\,(\text{LOC (REG (x), PREP* (y)))}$

where PREP* is a *region-generating* mapping

(3.c) $\text{PREP*} : D \to LR,$

associated with the prepositional concept PREP.[11] The notation introduced here shall be exemplified by the NP

>   the village at the river.

"The river" denotes the *reference object* (RO), which determines the frame of localization, "the village" refers to the object to be located (LO). The spatial relation between the two objects (on the level D)

(5.a)   AT (LO, RO)

is induced from a relation on the LR-level:

(5.b)   LOC (REG (LO), AT* (RO))

The general structure of the relation between the levels of objects and locations is described in fig. 4.

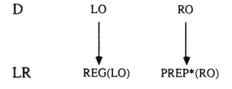

D   LO   RO

LR   REG(LO)   PREP*(RO)

Fig. 4: The relations between objects and their locations

(Remark: Because of the canonical type of the relationship between LO and REG(LO) I will sometimes use a relational notation as in (5.a) with respect to locations, i.e. objects of the domain LR. This usage is only a simplification, which always can be transformed into a fully precise expression of type (5.b).)

2.2 The case study of between

Now we will proceed with an analysis of between-PP's. I will start with some examples:

(6) a.   Coventry is between London and Liverpool.
     b.   Belgium is between France and the Netherlands.

In (6.a) the reference objects can be conceptualized as points (cp. fig 5.a), wheras in (6.b) spatially expanded ROs are mentioned. The possibility to conceputualize one object in one case as point in another case as area depends on the principles of cognitive granularity.[12] The further line of argumentation starts in section 2.2.1 with the most simple type of conceptualizations, namely points and straight lines, and goes on to section 2.2.3, where higher-dimensional localizations are described.

---

[11] This approach to the analysis of spatial prepositional phrases is a modification of Wunderlich / Herweg (to appear) and Herweg (1989). Habel / Pribbenow (1988) describe the concept of region-generating processes in detail.

[12] Compare the remarks on grain and size at the end of section 1 and the discussion of the example (23).

An important and characteristic property that distinguishes <u>between</u> from other prepositions is that the reference object in <u>between</u>-PPs is always a complex one described by a plural NP. The meaning of such complex NPs will be represented by means of a sum-operator of the propositional representation language (Cf. Link 1983). In sentences like

(6.c)   Myra and Barbara sat in the back, the baby between them... .[13]

a complex discourse object, here informally described by "Myra ⊕ Barbara", is introduced in the internal model (on the level of D of objects), and used as a reference object to locate the baby. (On this approach to plural discourse objects cf. Eschenbach et al. (1989).)

2.2.1 Points and lines

Problems of *dimensionality* are disregarded in the present paper. There are - at least - three important types of dimensionality problems which I will tackle in future research:

- mathematical problems, e.g.: What types of definition of *dimension* are relevant for cognitive space? Is it appropriate to use a fractal geometry approach, or are classical geometries sufficient?

- perception-oriented problems, e.g.: Are 2 1/2D-sketches (cp. Marr 1982) adequate only in perceptual processes, or are they also suitable for cognition, especially with respect to mental images?

- cognitive problems, e.g.: In which situations are projections for reducing dimensionality carried out? (Cf. example (6.a) in which towns are conceptualized as points.)

I will start with a one-dimensional basis, i.e. the orderings of points on straight lines.

In axiomatic geometry - following Hilbert (1977; 12th ed.) or Tarski (1959) - BETWEENNESS is usually seen as one of the basic relations. Without going into the details of a strict formalization of such a geometry, I will introduce a notation which is based on the Hilbert-approach, but shows some slight variations, e.g. infix-notation:

(7.a)   If a, b, c are different points of a straight line, then there exists exactly one of these points, say b, with: b BETWEEN <a, c>

Since BETWEEN is symmetric within its second argument,[14] i.e.

(7.b)   b BETWEEN <a, c> iff b BETWEEN <c, a>,

I will use in the following a set-oriented notation (7.c), which can be seen as a notational variant of (7,d).

---

[13] This sentence and some other examples in section 3 are from Collins Cobuild English Language Dictionary, 1987.

[14] The notions *symmetry, irreflexivity* and *transitivity* are used here in a loose manner; I transfer the core of their meaning from the classical case of two-place relations to the specific case of BETWEEN with a complex second argument, which means that BETWEEN is a kind of an at-least-three-place relation.

(7) c.   b   BETWEEN $\{a, c\}$

     d.   b   BETWEEN $(a \oplus c)$

Remark: Corresponding to the constraint of <u>between</u> that a plural NP is needed as a description of the reference object, there is a constraint on the level of semantic representation that the second argument of BETWEEN is a complex one, i.e. it is formed by summation (notation 7.d) or has cardinality more than 1 (notation 7.c).[15]

Some relevant properties of BETWEEN which will be needed in the following are:

(8.a)   BETWEEN   is irreflexive

           $\neg a$ BETWEEN $\{a, c\}$

           $\neg c$ BETWEEN $\{a, c\}$

(8.b)   BETWEEN   is transitive[16]

           if b   BETWEEN $\{a, c\}$ &   c   BETWEEN $\{b, d\}$

           then b   BETWEEN $\{a, d\}$ &   c   BETWEEN $\{a, d\}$

With

(9)   Let a, b, c be points on a line l.

     b   BETWEEN $\{a, c\}$ iff b $\in$ BETWEEN* $\{a, c\}$

we define a BETWEEN-region, which corresponds to a function BETWEEN*, inducing this region. For the basic situation discussed in the present section, the BETWEEN-region with respect to the points   a   and c   is the line segment induced by the points.

The BETWEEN-concept developed above is sufficient to reflect Moilanen's (1979) characterization of <u>zwischen</u> (<u>between</u>):

(10)   <u>zwischen</u> induces a linear relation. The reference objects restrict a line segment, on which the object whose location is to be indicated is situated. (Shortened translation from Moilanen 1979, p. 121 by C.H.).

Sentence (6.a) corresponds to a situation as described in (10); cf fig. 5.a. By a slight extension of the BETWEEN-concept from points on a line to line segments of one line, which is also proposed by Moilanen (1979), examples like (11) can be handled too (cf. fig. 5.b). The size of a mental map determines its resolution. Since the distance from Liverpool to Sheffield is relatively small, it is natural to conceptualize the three cities in (11) as areas and not as points.

(11)   Manchester lies between Liverpool and Sheffield.

---

[15] There are some strange exceptions to the plural-NP constraint, e.g. "Er saß zwischen dem Gerümpel" (He sat between the junk). An fully adequate explanation of these facts is not possible up to now. Furthermore, it is far from clear whether every complex object can be the RO of a BETWEEN-constellation.

[16] To be exact, the transitivity of BETWEEN is restricted to the contextual frame of one cognitive map. Transitivity conditions for spatial prepositions are described in detail in Habel / Pribbenow (1989).

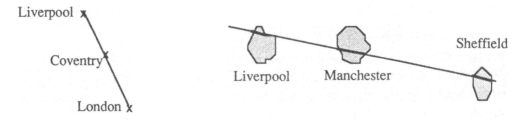

Fig. 5.a.: Line-based
between: lines and points

Fig. 5.b.:Line-based between: lines and areas

2.2.2 Paths

Careful consideration of further examples as

(12) a.   Duisburg lies between Essen and Düsseldorf
     b.   Tower Hill lies between the Monument and Liverpool St. Station

shows that in these cases there are no straight lines to constitute BETWEEN-regions. (A sketchy map with respect to (12.a) is fig. 6.a; fig. 6.b shows the relevant section of the map of the London Underground.)

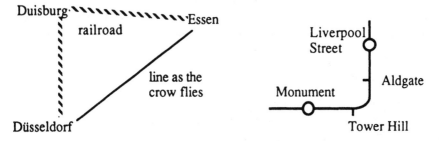

Fig. 6.a: German Intercity Railway

Fig. 6.b: London Underground

The use of <u>between</u> is permitted since there is a canonical *connection* from Essen to Düsseldorf or from Monument to Liverpool Street Station. More general, this means that beyond straight lines, further *reference paths* have to be taken into consideration for the formation of BETWEEN-regions.

Thus we have to investigate the following problems:
   - Which type of reference-path is suitable in which context/situation?
   - What is an adequate explication/formalization of the informal concept *reference path*?

I will begin with the second question. Similar to Wunderlich/Herweg (to appear) I use the *path* concept of topology (cf. on paths: Armstrong 1983):[17]

(13)   Let I denote the interval [0, 1] ⊂ **R**.
        Let LR be the set of local regions.[18]
        φ: I → LR is a *parametric path*, iff φ is continuous wrt the topology for LR.

Further relevant concepts have to do with the source and the goal of the path and with its set of points:

(14)   φ (0)   =: B (φ)   is called the *beginning point*
        φ(1)   =: E (φ)   is called the *end point*
        φ(I)   is called the *spur* (trace) of the parametric path φ.

It is obvious that these concepts are appropriate for distinguishing "the path from B to E" and "the path from E to B". This is needed especially in the analysis of local and directional concepts. Furthermore parametric paths give information about the *velocity of traversing* the spur from the beginning point to the end point. It is exactly this property of parametric paths which is the base of Bierwisch's (1988) criticism against the approach of Wunderlich/Herweg (to appear):

> "... We need a notion of path which is not based on time in the first place."
> Bierwisch, 1988, p. 15)

Without going into details, I now describe another concept of *path* that is more adequate for the analysis of most cases of <u>between,</u> based on equivalence classes of parametric paths:[19]

(15) a.   Let ψ: I → I.
           ψ is a *parameter-transformation* iff
           ψ is continuous, surjective and monotone increasing.
      b.   φ$_1$ and φ$_2$ are *equivalent parametric paths* iff
           there is a parameter-transformation ψ, such that:
           φ$_1$ = ψ ∘ φ$_2$, i.e. φ$_1$ (x) = φ$_2$ (ψ (x)).
      c.   An equivalence class Φ = [φ]$_ψ$ of parametric paths is called a *path*.

---

[17] Miller / Johnson-Laird (1976; p. 405ff) also use a similar path concept but without mentioning the correspondence to topology. Furthermore it is interesting that their paths are always based on the interpretation of the parameter as time variable. (See below the comments on Bierwisch vs. Wunderlich / Herweg.)

[18] LR, and this means the spatial representation format SPATREF, are assumed to make topological concepts available, especially the *neighbourhood* and *openness* of regions. In the present paper I will discuss neither the question as to which further topological concepts are needed, or how they are established (e.g. by a metric or pseudo-metric) nor I will go into the details of a topology on cell-matrices, which can be seen as a special case of *digital topology* (Rosenfeld, 1979).

[19] In mathematical topology, paths and parametric paths are often not distinguished. The reason of this uniform treatment is that in many cases the difference between these concepts is not relevant for mathematics. For example if a space is path-connected the existence of a PATH is necessary; obviously it is of no relevance whether PATH is seen as parametric path or as path, i.e. equivalence class of parametric paths.

A path neglects the path-traversing velocity of the parametrization but regards the orientation, given by beginning point and end point, and the spur.

This means that with respect to the formal description of path-like objects there is a triplet of concepts with decreasing content of information:

(16)

		velocity	orientation	location
parametric path	φ	$\varphi(x)$	$B(\varphi), E(\varphi)$	$\varphi(I)$
path	$\Phi = [\varphi]$	-	$B(\Phi), E(\Phi)$	$\Phi(I)$
spur	$Sp = [\Phi]$	-	-	$Sp(I)$

For the path-based analysis of BETWEEN a further definition is needed:

(17) A path Φ is *simple*, iff
$\forall x, y \in (0, 1) : x \neq y \Rightarrow \Phi(x) \neq \Phi(y)$
and $\forall x \in (0, 1) : \Phi(x) \neq \Phi(0)$ & $\Phi(x) \neq \Phi(1)$

A simple path has no point of its spur that is the image of two different points from the open interval (0,1). I.e. simpleness corresponds to the lack of "crossings (or intersections) in the spur".

With this additional concept I come back to sentence (12). The usage of <u>between</u>, e.g. in 12.b, is based on the existence of a path with the following properties

(18) $\Phi : I \rightarrow LR$
$\Phi(0) = $ REG (Monument)
$\Phi(1) = $ REG (Liverpool St. Station)
$\exists y \in (0, 1) : \Phi(y) = $ REG (Tower Hill)

This is a specific case of path-based BETWEEN-constellations:

(19) b BETWEEN $\{a, c\}$
iff there exists a simple path Φ with
$\Phi(0) = a$, $\Phi(1) = c$ and
$\exists y \in (0, 1)$ with $\Phi(y) = b$

(Note that straight lines are the spurs of specific paths, i.e. (19) is an extension of the characterization of section 2.1.)

For this path-based analysis of BETWEEN the relevant properties of irreflexivity and transitivity are preserved. These properties depend on the condition of simplicity (in case of irreflexivity) in (19) and the definition of summation (or concatenation) of paths (20) for transitivity.

(20) $\Phi_1 : [0, 1] \to LR$ $\Phi_2 : [0, 1] \to LR$
$\Phi_1(0) = a, \quad \Phi_1(1) = \Phi_2(0) = b, \quad \Phi_2(1) = c$
Then
$\Phi := \Phi_1 + \Phi_2$ defined by

$$\Phi(x) = \begin{cases} \Phi_1(2x) & x \in \left[0, \frac{1}{2}\right) \\ \Phi_2(2x-1) & x \in \left[\frac{1}{2}, 1\right] \end{cases}$$ is called the sum of Φ_1 and Φ_2

Furthermore: $\Phi(0) = \Phi_1(0) = a$, $\Phi(1) = \Phi_2(1) = c$

It is obvious that the characterization (19) is too general, since in the plane every constellation consisting of three points a, b and c can be connected by a path Φ from a to c with b on it. This means that the existence of a connecting path is too weak as defining conditions of BETWEEN-regions. In other words, only some specific paths are capable of inducing a BETWEEN-region. In the present paper, I can only give a sketchy overview on some types of *natural paths* between two objects.

- straight lines: these are specific spurs which can be constructed independent of any context or situation.

- traffic connections, such as railways, roads, rivers, ...

- natural pathlike objects, such as coastlines, brooks, edges and ridges of mountains, ...

- induced paths, such as lines of telegraph poles,...

The second and third type of paths correspond to psychological concepts known from the work of Lynch (1960). His work on mental maps of cities shows that *paths*, which are the entities humans move on when they go from one point to another, and *edges*, which are boundary lines that separate *districts* from each other, have great importance (cf. Miller/Johnson-Laird, 1976, 377-378). Note that this dichotomy of connecting and separating is also fundamental in topology. This is particularly evident in the formal notion of *connectivity* of a topological space. The more specific concept of path-connectivity depends on the existence of paths between points of the topological space. The more general concept of connectivity is based on the non-existence of partitions, i.e. on the impossibility of separating points or subsets of the topological space. (Cp. on connectivity: Sutherland 1975).

The *district* concept of Lynch (1960) as well as *induced paths* (see above) demonstrate that principles of Gestalt psychology are relevant for the constitution of path-like objects (Cf. Attneave (1954) and Koffka (1935)). From a topological point of view, the *boundary* of a district corresponds to the *frontier* of a set (Armstrong 1983).

Since there can be more than one natural connecting path from A to B, it is possible that more than one BETWEEN-region wrt A $\oplus$ B exist. This "ambiguity" is exemplified by

(21) a. Tower Hill lies between Monument and Liverpool Station.

b. St. Peter-Upon-Cornhill lies between the Monument and Liverpool St.

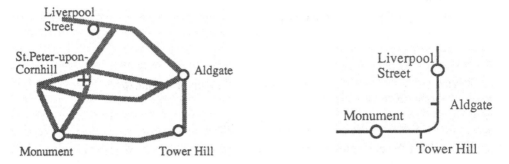

Fig. 7.a: "From Monument to Liverpool Street Station"   Fig. 7.b: London Underground

Monument and Liverpool St. can be connected in two ways: by the underground (cf. fig. 7.b) or on ground via streets (fig. 7.a). (Note that the connection by streets is nearly a straight line here.) Background knowledge about underground stations triggers access to the appropriate knowledge entities. Whether it is necessary to mention the type of the connecting path explicitly or not, e.g. to extend (21.a) with the modification "on Circle Line", depends on assumptions about the knowledge shared by speaker and listener.

These example (21) and figures 7 are interesting with respect to further phenomena of and questions about mental maps. How many mental maps are used, and how relations among mental maps have to be understood, are open questions at present. The two connections between Monument and Liverpool St. are supposed to be parts of different mental maps. Under this assumption, the search for a connecting path is limited to specific maps; thus BETWEEN-regions are generated with respect to specific maps as well. (I will come back to this below.)

Furthermore, the sketchy map of London Transport is geometrically different from the real world topography; but there exists a topological embedding from the underground world (fig. 7.b) into the surface world (fig. 7.a). The existence of such embedding and other connections among different mental maps give evidence against the supposition that traffic connections are always conceptualized as straight lines.[20]

Now it is time to come back to the question of whether between-predication of the type discussed in this section are *path-based* or *spur-based*. I assume that for most cases, the spur is sufficient, and paths are only needed for inducing the spur. This relationship between a path and its spur can be traced back to the kinetic experience of individuals, which leads to the knowledge about connections (cp. Briggs, 1973). But is it always

---

[20] Traditional proposals for the analysis (or explanation) of between-phrases follow usually a straight-line-strategy; as evidence for this type of conceptualization the straight-line-sketches in the subway trains are often mentioned. I believe that this assumption is not well founded.

sufficient to use spurs for generating and comprehending between-constructions? Suppose the following sentences are uttered in a train of the Circle Line at King's Cross going clockwise:

(22)   To get to Tower Hill, you have to leave the train
   a. between Liverpool St. and Monument.
   b. ?between Monument and Liverpool St.

The sentence b is unacceptable because of the opposition in direction of "from Monument to Liverpool St." and "clockwise on the Circle Line". This situation is not a counter-example to the spur-based analysis of between; the path, i.e. the orientation is relevant in (22), since a situational origo, namely King's Cross, and situation-dependent orientation, induced by a train, come into play.

I will now come back to the question of systems of cognitive maps. Whereas it is appropriate to say (23.a), it is not appropriate to localize a specific building of New York in the BETWEEN-region in question by (23.b), (cf. fig. 8).

(23) a. New York lies between Hamburg and Los Angeles.
      b. Empire State Building is between Hamburg and Los Angeles.

This means that rules like

(24)   If b BETWEEN {a, c}
        and d IN b, i.e. LOC(REG (d), REG (b)),
        then d BETWEEN {a,c}

are not always applicable[21]. Without going into the details here, I will give a sketch of an explanation. The BETWEEN-region with respect to Hamburg and Los Angeles is formed in a mental map corresponding to the sketchy map of fig. 8.

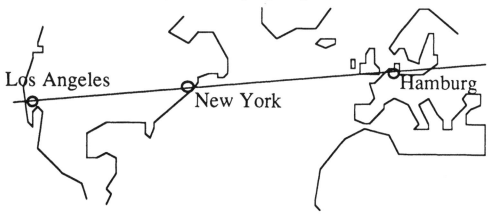

Fig. 8: "between Hamburg and Los Angeles"

---

[21] Similar problems, especially those of the transitivity of local relations, are discussed in detail in Habel / Pribbenow (1989). In constellations as mentioned in sentence (23.b) some Gricean maxim is violated by using a inference rule like (24).

One of the fundamental properties of mental maps concerns their granularity, which depends on and corresponds to the size of the map (cf. section 1). Now it is intuitively obvious that Empire State Building is too small to become part of a map spanning from California to Europe just because of *granularity*. (On the general concept of *granularity* - beyond the spatial domain - cf. Hobbs, 1985.) The conceptualization of one object in different mental maps corresponds slightly to Kautz' (1985) approach, in which one object is localized with respect to different *grains*. (24) demonstrates that the inferences which are based on semantic relations among spatial concepts are strongly dependent on the mental map (or grain) of the objects in question, i.e. inferences between different mental maps are not possible in general.

2.2.3 Higher-dimensional constellations

Up to now I only have discussed situations in which one path (or spur) constituting an appropriate connection from A to B is considered, to decide whether C belongs to BETWEEN*{A, B}. In other words, since paths intuitively can be seen as one-dimensional spatial objects, the BETWEEN-regions considered in the previous section are one-dimensional or induced by distinguished one-dimensional regions, namely the *natural paths*.

At this point of discussion let us look at this sentence

(25)   Vechta lies between Osnabrück and Bremen.

The spatial constellation is depicted in fig. 9. The problem in this situation is that none of the canonical paths which connects Osnabrück with Bremen passes through Vechta, neither the straight line Φ_1 nor the railway Φ_2 nor the motorway Φ_3.

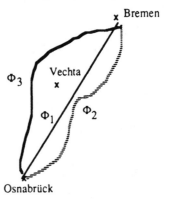

Fig. 9: Deformation of paths

Since paths are topological concepts it is suitable to use topological concepts to explain situations as those of (25) and fig. 9. The following proposal is based on a specific relation between paths (the canonical connections), namely the *homotopy* relation, which is induced by continuous *deformation*.

(26)   Let φ_1 and φ_2 be parametric paths with $\varphi_1(0) = \varphi_2(0)$ and $\varphi_1(1) = \varphi_2(1)$,
        and let $\delta: I \times I \to LR$ be a map with
        $$\delta(0, y) = \varphi_1(y) \qquad\qquad \delta(1, y) = \varphi_2(y)$$
        $$\delta(x, 0) = \varphi_1(0) = \varphi_2(0) \qquad \delta(x, 1) = \varphi_1(1) = \varphi_2(1)$$
        Then δ is called a deformation and
        φ_1 and φ_2 are deformable into (are homotopic to) each other.

Without going to the details, it can be stated that φ_1, φ_2 and φ_3 (from fig. 9) are homotopic to each other. Based on this concept of deformation I propose a characterization for two-dimensional BETWEEN-regions.

(27)   If Φ_1 and Φ_2 are homotopic paths from A to B
         and δ a deformation which induces the homotopy,
         then for all C in the range of the deformation δ,
         i.e. $\exists\, x \in [0, 1], y \in (0, 1) : \delta\,(x, y) = C$,
         holds $C \in$ BETWEEN*$_{\Phi_1, \Phi_2}$ {A, B}

The intuitive idea behind (27) is that during the deformation from φ_1 to φ_2, i.e. in the process of adjusting one path to the other, the point C is on one of the paths in between φ_1 and φ_2. In (27) I use the paths as indices to the region-constituting function to indicate that this BETWEEN-region is dependent on these connecting paths. The definition (27) leads the direction for further research to be done in the future:

- a concept of deformation that is not restricted to two-dimensional space, i.e. that the idea of BETWEEN-regions as regions passed during deformation is suitable for three-dimensional constellations, too.

- analogous to the naturalness condition of connecting paths, it is necessary to formulate naturalness conditions for deformations. Such conditions seem to depend on metric properties.

- BETWEEN-regions can also be induced by deformations of one path only. Fig. 10 depicts a situation in which the sentence

(28)   Annapolis lies between Washington and Philadelphia.

    can be interpreted. It is irrelevant which of the canonical paths between Washington and Philadelphia (the straight line or the interstate) are used as basis of the deformation.

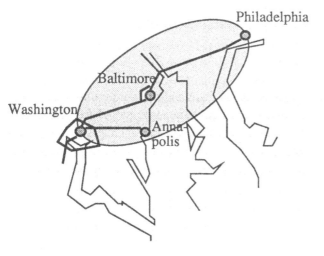

Fig. 10: Deformation of one path

At this point I want to end the detailed exposition and continue with an overview. A straightforward extension of the path-based analysis explained above is the use of convex hulls. The convex hull of a set C is the intersection of all convex sets containing C (wrt an *affine space*). Thus, if two points are in the convex hull, then so are all the points on the line segment joining them. (This defining property of *convex sets* justifies seeing convex hulls as extensions of paths.) This extension is (cf. Habel 1989) appropriate for describing and explaining <u>between</u>-expressions with more than two reference objects or with higher-dimensional reference objects. For example, in sentence (29) the convex hull of the two matches is the BETWEEN*-region of the RO in question (Cf. fig 11).

(29)   The button lies between the matches.

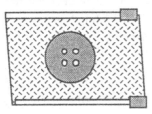

Fig. 11: the button between the matches

To summarize the argument:

- cognitive space, realized over SPATREF, possesses a topological structure. The evidence of <u>between</u>-expressions (in section 2) and of <u>um</u> (example (1) and fig. 3) shows that the *path* concept is relevant for natural language analysis.

- it is necessary to formulate naturalness conditions for connecting paths and deformation. Such conditions depend not only on prior (background) knowledge but also on metric constraints. This means that LR has to be *metric* or *pseudo-metric*.

With respect to non-spatial <u>between</u> it is interesting that a wide spectrum of phenomena can be explained by paths with respect to linear orderings (Habel 1989)

3. "Follow me!" to the summary

The investigation of <u>between</u> has demonstrated that paths and spurs are useful concepts for the analysis of spatial expressions. What about with parametric paths? Remember that Bierwisch's (1988) criticism against Wunderlich/Herweg (to appear) refers to exactly this point. In the present section I will only sketch of the use of paths and spurs in the analysis of one specific verb, namely <u>follow</u>.[22] The most easy case to handle - but that one, which

---

[22] Similar considerations are described in Hays (1990). Her research is based on the concept *trajectory*, which is widely used in the area of Computer Vision. The explicit distinction of different path concepts, which is summarized in (16) has no counterpart in Hay's use of trajectories.

is most distant from the canonical sense of <u>follow</u> - concerns situations with two static discourse objects, namely subject and object:

(30)   The boundary follows the Rio Grande.

Since the objects in question have the semantic (or conceptual) property [+ static], only spurs can be referred to.[23] It is obvious that the spurs of the two objects have to be in a tolerable distance, i.e. are deformable to each other. (Note, that naturalness conditions with respect to deformation are needed here, too).

The next, more complicated situation concerns a static object and a dynamic subject, i.e. a static RO and a dynamic LO, as in

(31)   We follow up a path along the creek ...

The spur of the follower's path "follows the spur" of the followed object, in this case the creek, in just the same way and with the same constraints as in the case of (30). Up to now, nothing is different from of <u>between</u>; only spurs and paths are needed. But consider

(32)   a.   He followed them to Venice.
      b.   The detective followed the man with the black coat.

Here we need parametric paths, since we have to take some constraints into consideration:

-   There are constraints with respect to time and space, e.g.
    $$\forall\, t \in [0, 1] : \varphi_1(t) \ \text{BEHIND} \ \varphi_2(t)$$
    w.r.t.   a tolerance measure.

-   It remains open what distances between the spurs and what intervals between the time slices $\varphi_1(t)$ and $\varphi_2(t)$ are tolerable.

The three types of <u>follow</u>-uses (as also described by Hays, 1990)

-   both subject and object (LO and RO) are stationary,
-   LO is moving, RO is stationary,
-   both are moving

are distinguishable with respect to the types of path-concepts that are needed from the parametric path - spur triple (16). But there are some complications in the simple situation I just described:

-   Why is the case "LO stationary and RO moving" missing?

Note that (33) is not a counterexample to the assumption that such cases are lacking.

(33)   Eric and the others could follow the man's finger as it moved across the map.

---

[23]Hays characterizes such examples as "strictly static use", which "can also be seen as a metaphoric extension from the canonical sense." I will come back to this point later.

In (33) the focus of vision is the induced moving object which follows the moving RO. Analogous to this, example (30) can be explained by assuming a process of scanning the "boundary following the Rio Grande" in a mental image.[24] Further open problems are:

- Is the situation of (30), a situation consisting of two stationary objects, symmetric? In other words, are LO and RO changeable?
- Can any pair of objects with a path-like shape be situated in a <u>follow</u> situation? For example, why is it not possible to describe a situation as depicted in fig. 11 by "one match follows the other."? (Here the strictly local use of <u>follow</u> as a position verb is meant. The inherent direction of the matches is relevant.)

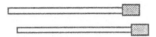

Fig. 12: *One match follows the other

One possible explanation of these facts is that there must be a conceptualization of LO, the subject, which induces a moving (or movable) object[25], e.g. the moving focus of vision in scanning a mental image. In other words, if "LO follows RO", it is necessary that the subject (LO) is "at least as movable with respect to the FOLLOW-situation" as the object (RO). In the match example (fig 12) it is possible that grain and size of the image are important; the objects in question are so small that scanning along the object's shape is unnatural.

To sum up, <u>follow</u> is based on parametric paths, which in specific cases can be reduced to paths or spurs. The acceptable LO-RO-constellations are constrained by the appropriate conceptualizations.

I will conclude this section with a summary and some remarks on open questions. These cases of <u>between</u> (as a representative of some spatial prepositions) and <u>follow</u> (as a representative of verbs of motion) have shown that the path concept is fundamental to the description and explanation of spatial concepts of natural languages. The analysis has demonstrated that topological, metrical, and (with fewer arguments in the present paper) geometrical concepts are relevant to cognitive models of spatial constellations; LR, the domain of local regions, has to be seen as a space with topological, metrical, and geometrical structure.

This line of argumentation and explanation does not strictly follow traditional methods of mathematics. Some stumbling blocks are based on the finiteness and the graded granularity of mental images. These phenomena lead to special problems (in contrast to "pure

---

[24] A similar explanation of the facts is possible for the PP "entlang dem Rio Grande" (along the Rio Grande). On the relation between motion verbs and so called "path prepositions" see also Retz-Schmidt (1988).
[25] "Movable" here refers to a very general concept, which also can involve changes of form or size (cp. Miller / Johnson-Laird 1976, p.529).

mathematics") and specific solutions. For example, principles of the embedding of mental maps, the different instances of LR, have to be investigated.

This summary can be summarized as follows: A formal theory of cognitive processes in spatial reasoning will be based on the mathematical theories of topological, metrical and geometrical spaces, but has to respect the specific constraints of finite size and limited granularity of local regions in mental models.

References

Armstrong, M. A. (1983): Basic Topology. Springer: New York.

Attneave, F. (1954): Some informational aspects of visual perception. Psychological Review 61. 183-193.

Bierwisch, M. (1988): On the grammar of local prepositions. In: M. Bierwisch / W. Motsch / I. Zimmermann (Hrsg.) Syntax, Semantik und Lexikon. Studia Grammatica 29. Akademie-Verlag: Berlin. 1-65.

Blank, A. (1958): Axiomatics of binocular vision: the foundations of metric geometry in relation to space perception. Journal of the Optical Society of America 48. 328-333.

Block, N. (ed.) (1981): Imagery. MIT-Press: Cambridge, Mass..

Briggs, R. (1973): Urban cognitive distance. in: R. Downs & D. Stea (eds.): Image and Environment. Aldine Publ.: Chicago. 361-388.

Codd, E. F. (1968): Cellular Automata. Academic Press: New York.

Downs, R. / Stea, D. (1977): Maps in Minds. Harper & Row: New York.

Eschenbach, C. / Habel, Ch. / Herweg, M. / Rehkämper, K. (1989): Remarks on Plural Anaphora. in: Proc. 4th Conf. European Chapt. of the ACL. 161-167.

Habel, Ch. (1986): Prinzipien der Referentialität. Springer: Berlin.

Habel, Ch. (1987): Cognitive Linguistics: The Processing of Spatial Concepts. in T.A. Informations -ATALA (Association pur le Traitement Automatique des Langues .28. 21-56 (also as: IBM: Stuttgart. LILOG-Report 45, 1988)

Habel, Ch. (1989): zwischen-Bericht. in: Ch. Habel / M. Herweg / K. Rehkämper (Hrsg.): Raumkonzepte in Verstehensprozessen. Niemeyer: Tübingen. 1989. 37-69.

Habel, Ch. / Pribbenow, S. (1988): Gebietskonstituierende Prozesse. IBM: Stuttgart. LILOG-Report 18.

Habel, Ch. / Pribbenow, S. (1989): Zum Verstehen räumlicher Ausdrücke des Deutschen - Transitivität räumlicher Relationen in: W. Brauer / C. Freksa (Hrsg.); Wissensbasierte Systeme. Informatik Fachberichte. Springer: Berlin. 139-152.

Hays, E. M. (1990): On defining motion verbs and spatial prepositions. to appear in: Ch. Freksa / Ch. Habel (Hrsg.): Repräsentation und Verarbeitung räumlichen Wissens. Springer: Berlin.

Herweg, M. (1989): Ansätze zu einer semantischen Beschreibung topologischer Präpositionen. in: Ch. Habel / M. Herweg / K. Rehkämper (Hrsg.): Raumkonzepte in Verstehensprozessen. Niemeyer: Tübingen. 99-127.

Hilbert, D. (1977, 12th ed.): Grundlagen der Geometrie. Teubner: Stuttgart.

Hobbs, J. R. (1985): Granularity. Proc. 9th IJCAI. 432-435.

Johnson-Laird, P.N. (1983): Mental Models. Cambridge UP: Cambridge.

Kautz, H. (1985): Formalizing spatial concepts and spatial language. in Hobbs, Jerry et. al.: Commonsense summer: Final report. CSLI, Stanford Cal. Report No. CSLI-85-35. 2-1 - 2-45.

Khenkhar, M. (1988): Vorüberlegungen zur depiktionalen Repräsentation räumlichen Wissens. IBM: Stuttgart. LILOG-Report 19.

Koffka, K. (1935): Principles of Gestalt Psychology. Harcourt Brace: New York.

Kosslyn, S. (1980): Image and Mind. Harvard UP: Cambridge, Mass..

Link, G. (1983): The logical analysis of plurals and mass terms: a lattice-theoretical approach . in: R. Bäuerle/Ch. Schwarze/A. von Stechow (Hrsg.): Meaning, Use, and Interpretation of Language. de Gruyter: Berlin. 302-323.

Lynch, K. (1960): The Image of the City. MIT-Press: Cambridge, Mass..

Marr, D. (1982): Vision. Freeman: New York.

McNamara, T. (1986): Mental Representations of Spatial Relations. Cognitive Psychology 18. 87-121.

Miller, G. / Johnson-Laird, P.N. (1976): Language and Perception. Cambridge Univ. Press. Cambridge.

Paivio, A. (1983): The Empirical Case for Dual Coding. in: J. Yuille (ed.): Imagery, Memory and Cognition. Erlbaum: Hillsdale, N.J.. 307-332.

Pylyshyn, Z. (1981): The Imagery Debate: Analogue Media versus Tacit Knowledge. Psychological Review 88. 16-45.

Rehkämper, K. (1988): Mentale Bilder - Analoge Repräsentationen. IBM: Stuttgart. LILOG-Report 65. also to appear in: Ch. Freksa / Ch. Habel (Hrsg.): Repräsentation und Verarbeitung räumlichen Wissens. Springer: Berlin.1990.

Retz-Schmidt, Gudula (1988): Various views on spatial prepositions. AI Magazine 9. 95-105.

Roberts, F. S. (1973): Tolerance geometry. Notre Dame Journal of Formal Logic 14. 68-76.

Roberts, F. S. / Suppes, P. (1967): Some problems in the geometry of visual perception. Synthese 17. 173-201.

Rosenfeld, A. (1979): Digital topology. American Mathematical Monthly 86. 621-630.

Sutherland, W. A. (1975): Introduction to Metric and Topological Spaces. Oxford University Press: Oxford.

Tarski, A. (1959): What is elementary geometry? in: L. Henkin / P. Suppes / A. Tarski (eds). The axiomatic method, with special reference to geometry and physics. North-Holland Publ.: Amsterdam. 16-29.

Westheimer, G. (1987): Visual Space. in Gregory, Richard L. (ed.): The Oxford Companion to the Mind. Oxford University Press: Oxford. 796.

Wunderlich, D. / Herweg, M. (to appear): Lokale und Direktionale. in: A. v. Stechow / D. Wunderlich (eds.): Handbuch der Semantik. deGruyter: Berlin.

SLOT GRAMMAR
A System for Simpler Construction of
Practical Natural Language Grammars

Michael C. McCord
IBM Thomas J. Watson Research Center
P. O. Box 704
Yorktown Heights, NY 10598

Abstract

Slot Grammar makes it easier to write practical, broad-coverage natural language grammars, for the following reasons. (a) The system has a lexicalist character; although there are grammar rules, they are fewer in number and simpler because analysis is largely data-driven − through use of *slots* taken from lexical entries. (b) There is a modular treatment of different grammatical phenomena through different rule types, for instance rule types for expressing linear ordering constraints. This modularity also reduces the differences between the Slot Grammars of different languages. (c) Several grammatical phenomena, such as coordination and extraposition, are treated mainly in a language-independent *shell* provided with the system.

1 Introduction

The Slot Grammar system provides a convenient means for writing practical, broad-coverage grammars for natural languages. The grammar for a given language is simplified, and the *differences* between grammars of different languages are reduced, for the following reasons.

(1) Slot Grammar has a highly *lexicalist* character. Grammatical analysis makes systematic use of *slots* (essentially syntactic relations) obtained from lexical entries. Although there are grammar rules, they tend to be fewer in number and simpler because the system is more data-driven.[1]

(2) There is a modular treatment of grammatical phenomena through different rule types. In particular, linear order is treated modularly by separate *slot ordering rules*. This not only

---

[1]   This does not mean that complexity is just shoved to the lexicon. Mainly, it means making better use of lexical information, information that is needed anyway for applications of the grammar such as machine translation.

simplifies a given grammar, but also reduces differences across grammars because linear order is one of the main ways in which languages differ.

(3) The Slot Grammar system has a language-independent *shell*, which includes not only the parser but also a treatment of several grammatical phenomena specific to natural language. In particular, the shell contains much of the treatment of coordination, left extraposition (left movement) and remote dependencies, implicit subjects, and punctuation. Of course there are aspects of these phenomena that depend on the particular natural language, so the general procedures are "parametrized" by calls to simple rules (mainly unit clauses) in specific grammars. One way of viewing the shell is that it makes the Slot Grammar system a higher-level, special-purpose system for writing natural language grammars. Unlike some augmented phrase structure grammar systems (such as DCGs), Slot Grammar would be inappropriate for the grammar of, say, computer programming languages. But its special-purpose features offer greater simplicity and compactness for natural language grammar.

(4) The shell contains a parse evaluation system, extending Heidorn's [1982] parse metric, which not only ranks final parses but is also used for pruning away unlikely partial analyses during parsing. (The parser is a bottom-up chart parser.) The Slot Grammar parse evaluator effects weighted preferences for complements over adjuncts and for parallelism in coordination, as well as for close attachment. The parse space pruning system contributes to the simplification of Slot Grammars for particular languages because it offers language-universal constraints which for some grammatical systems would go in the language-specific grammar.

The original work on Slot Grammar was done in 1976-78 and appeared in [McCord 1980]. This earlier system was developed without consideration of logic programming (and was implemented in Lisp). The logic-programming-based grammatical systems described in [McCord 1982, 1985, 1987] represented a *combination* of the original Slot Grammar techniques with the (augmented) phrase structure grammar techniques common in logic programming [Colmerauer 1978]. In this combined approach, Slot Grammar rules, expressed in terms of phrase structure rules and Prolog clauses, were used systematically for postmodification of open-class words; but elsewhere in the grammar, more standard phrase structure rules were used. In particular, the combined approach was used in the Modular Logic Grammar **ModL** [McCord 1985, 1986, 1989a].

Work on the current Slot Grammar system began in 1988. The idea has been to develop the original lexicalist, head-driven approach into a practical system taking good advantage of logic programming. The main motivation for the new version has been its use in the machine translation system **LMT** [McCord 1986, 1989a,c,d, McCord and Wolff 1988]. **LMT** originally used Modular Logic Grammar for source analysis, but has been revised to use Slot Grammar. This has been extremely useful in making **LMT** capable of dealing with multiple language pairs – because of the Slot Grammar shell and the lexicalism.

Slot Grammar has some features in common with currently popular grammatical systems such as LFG, FUG and HPSG (see Shieber 1986 for an overview), in which there are themes of lexicalism and use of unification, and sometimes a dependency (head-driven) orientation. The original (1976-78) Slot Grammar work was done independently of these lines of work. The current Slot Grammar system differs from these related systems in several ways that will become apparent below. One general remark, though, is that (current) Slot Grammar could be said to be closer to Prolog — e.g. it uses term unification instead of attribute-value system unification. This may allow Slot Grammars to be more efficient and practical. And it does not entail a total sacrifice of formal neatness and purity, because the grammar rules can often be written declaratively (translating into (pure) Horn clauses).

For more comparison of the new Slot Grammar with related systems and with the earlier Slot Grammar system, see [McCord 1989b].

Slot Grammars for English, German, and Danish are being written respectively by the author, Ulrike Schwall, and Arendse Bernth, and are being used in versions of **LMT** with these source languages.

The output of Slot Grammar analysis is a syntax tree that shows both surface structure and deep grammatical relations. Such trees are the input to transfer in **LMT**. However, these trees are also quite amenable to deeper semantic analysis, partly because they already contain the predicate-argument structure for open-class words. In fact, there is a semantic interpreter which operates on the syntax trees and produces logical forms in the semantic representation language **LFL** [McCord 1987]. This in turn is input to the discourse understanding system **LODUS** of Bernth [1989, 1990].

The Slot Grammar output also seems quite amenable to the formulation of algorithms for syntactic constraints on anaphora [Lappin and McCord 1990]. Lappin is also working with the author on VP anaphora in the framework of Slot Grammar.

Section 2 of the paper describes what is accomplished with Slot Grammar analysis, in terms of its input and output. Section 3 discusses the lexical analysis phase of Slot Grammar analysis, which precedes syntactic analysis in a first pass. Section 4 describes the basic ingredients (rule types) of the Slot Grammar of a specific language. Sections 5, 6 and 7 are devoted respectively to extraposition, coordination, and implicit subjects (which are treated largely in the shell). The main algorithm for syntactic analysis is sketched in Section 8. Finally, Section 9 contains a description of parse evaluation and pruning.

2 Input and output of Slot Grammar analysis

Input consists of unrestricted text.[2] The input is segmented by a "sentence" separator, and these segments are analyzed one at a time. The segments need not be complete sentences. (In the following, the input segments will usually be referred to as "sentences" for convenience, however.)

The output of Slot Grammar analysis of a sentence is a parse tree which shows both surface structure and deep grammatical relations, including remote dependencies, in a single structure. An example is shown in Figure 1.

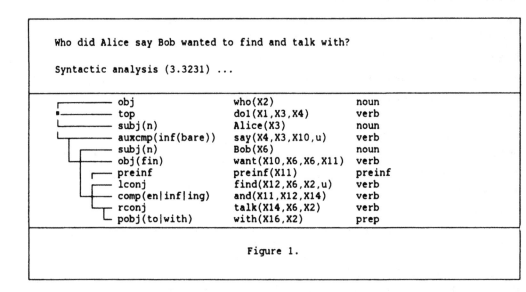

```
Who did Alice say Bob wanted to find and talk with?

Syntactic analysis (3.3231) ...

┌─────── obj            who(X2)              noun
■─────── top            do1(X1,X3,X4)        verb
└─────── subj(n)        Alice(X3)            noun
 └────── auxcmp(inf(bare)) say(X4,X3,X10,u)  verb
    ─── subj(n)         Bob(X6)              noun
    ─── obj(fin)        want(X10,X6,X6,X11)  verb
   ┌─── preinf          preinf(X11)          preinf
   ─── lconj            find(X12,X6,X2,u)    verb
  └─── comp(en|inf|ing) and(X11,X12,X14)     verb
   ┌─ rconj             talk(X14,X6,X2)      verb
   └─ pobj(to|with)     with(X16,X2)         prep
```

Figure 1.

The parse tree is a kind of dependency tree; every node has a head word. In the display, there is only one line per node, showing (1) the tree connection lines, (2) the slot filled by the node (phrase), (3) the head *word sense predication*, i.e., a sense of the head word, together with its arguments, and (4) the feature structure of the node. In Figure 1, the feature structure have been abbreviated, by a display option, to their principal functors (normally the part of speech of the head word).

The first argument of the word sense predication is the *marker variable* for the node. For verbs, this is like an event or state variable; for nouns it represents the main entity referred to by the noun. The other arguments correspond (in the main) to complement slots associated with the head word. For example, the second argument of a verb sense predication corresponds to the verb's (logical) subject. These complement argument variables are unified, through slot filling, with the marker variables of the filler phrases. Note in Figure 1 that *Who*, with marker x2, is

---

[2] There is a partial treatment of text-formatting tags for SCRIPT/GML which allows LMT to translate (some) SCRIPT source text for the source language into SCRIPT source for the target language.

shown as the object of both *find* and *with*, and *Bob*, with marker X6, is shown as the subject of *want, find,* and *talk.*

Internally, a syntax tree (node) is represented by a Prolog term

```
phrase(X,H,Sense,Features,SlotFrame,Ext,Mods),
```

where the components are as follows: (1) X is the marker variable of the phrase. (2) H is an integer representing the position of the head word of the phrase. (3) Sense is the word sense of the head word. (4) Features is the feature structure of the head word *and* of the phrase. It is a logic term (not an attribute-value list), which is generally rather sparse in information, showing mainly the part of speech and inflectional features of the head word. (5) SlotFrame is the list of complement slots, each slot being in the internal form slot(Slot,Ob,X), where Slot is the slot name, Ob shows whether it is an obligatory form of Slot, and X is the *slot marker.* The slot marker is unified (essentially) with the marker of the filler phrase when the slot is filled, even remotely, as in left movement or coordination. (6) Ext is the list of slots that have been *extraposed* or *raised* to the level of the current phrase. (7) The last component Mods represents the modifiers (daughters) of the phrase, and is of the form mods(LMods,RMods) where LMods and RMods are the lists of left modifiers and right modifiers, respectively. Each member of a modifier list is of the form Slot:Phrase where Slot is a slot and Phrase is a phrase which fills Slot. Modifier lists reflect surface order, and a given slot may appear more than once (if it is an adjunct). Thus modifier lists are not attribute-value lists.

Parse space pruning can be turned on or off dynamically by the user. When it is off, all (final) analyses are shown, ranked best-first by the parse evaluator. However, for long sentences the system may have a space overflow because of the large number of partial analyses in parsing (when pruning is off). When pruning is on, the final number of parses is very small – averaging about 1.5 per sentence. In spite of the drastic action of pruning away partial results during parsing, the system seems to do well at producing a correct result among the final results, modulo some reattachments of phrases such as PPs.

The parser always produces an output tree even when the input is not grammatical (for the given grammar). In such cases, the output is a "fitted" parse roughly similar to those produced by the PLNLP system [Jensen and Heidorn 1982]. **LMT** can use these fitted parses to produce a translation (sometimes good, sometimes bad!) for non-grammatical inputs.

There is a fairly complete system for handling files of sentences in either batch or interactive mode.

3 The lexical analysis phase

After sentence separation and tokenization have operated, the lexical analyzer operates on the words (tokens) of the sentence to produce *word frame* clauses which will be used by the syntactic analyzer (the main part of the parser).

A word frame clause represents a lexical analysis of a particular word in the sentence, specifying a sense, a feature structure (showing part of speech and inflection mainly), and a slot frame. Of course a word can have several analyses, so there may be several word frame clauses associated with one word. However, different occurrences of the same word may have different collections of analyses, because there are heuristics that remove some possible analyses, based on the local lexical context. For this reason, a word (token) is referred to in a word frame clause by its numerical position in the sentence. A word frame clause is then a unit (unconditional) clause of the form

```
wframe(WordNo,Sense,Features,SlotFrame).
```

The four arguments of `wframe` are essentially the same as the second through fifth arguments of a `phrase` structure, as described in the preceding section. In fact, the bottom-up parser begins its work by constructing one-word `phrase` structures from `wframe` clauses, in the obvious way – namely by adding empty lists of modifiers and extraposed slots.

As an example of a word frame, one `wframe` clause for the word *given* in the sentence *Alice has given Bob a book* might be:

```
wframe(3,s(give,1),verb(pastparta),
       slot(subj(n),op,X).
       slot(obj,op,Y).
       slot(iobj,op,Z).nil).
```

The sense name argument is generated automatically by the system. The feature structure argument shows that the word is the active past participle form of a verb.

Other word frame clauses could be generated for the word *given*. For example, for the sentence *The book was given to Bob*, a *passive* past participle frame would be appropriate:

```
wframe(4,s(give,1),verb(pastpart),
       slot(agent,op,Y).
       slot(subj(n),op,X).
       slot(iobj,op,Z).nil).
```

Lexical analysis actually involves two passes over the words of the sentence: (1) morpholexical analysis and (2) lexical filtering. The first pass produces a list of word frames for each (derived or inflected) word in the sentence and the second pass erases (filters out) some of these analyses, based on lexical context.

The first pass itself involves two steps for each word. In the first step, morphological analysis (affix stripping and consultation of tables of irregular forms) is done in conjunction with look up in lexicons stored in *external forms*, to be discussed below. This step produces derived analyses which are still basically in external form. Then a system for *lexical compiling* produces the wframe clauses (internal form) associated with these derived analyses. This system for morpholexical analysis is much as is described in preceding papers [McCord and Wolff 1988, McCord 1989a]. Much of the code for morpholexical analysis is in the shell.

In the lexical filtering pass, multiword lexical entries are used to replace sequences of adjacent words by single, conglomerate "super words" (which receive a single word number in word frame clauses). In addition, *lexical erasure rules* are applied to erase some word frame analyses based on surrounding analyses (useful for source languages like English with lots of part-of-speech ambiguity). For example, an active past participle frame, like the first one for *given* above, might be erased if there is no occurrence of a perfect *have* in the sentence. Lexical erasure rules are included in the Slot Grammar **XSG** of a specific language. There is a convenient rule formalism for them, and a rule compiler in the shell, but they will not be described further here due to lack of space.

Lexicons

There is a *standard external form* for lexicons useable by Slot Grammars. In this format, a lexical entry for a word Word is represented by a clause

    Word < Analysis.

where the Analysis shows word frame information in a convenient, abbreviated, external form. For example, an entry for *give* might be simply

    give < v(obj.iobj).

The subject slot is omitted here because by default it is added by the lexical compiler. This external lexical form is essentially the same as is described in earlier papers on **LMT**.[3]

Two lexicons used by **ESG** (the English Slot Grammar) are the following. There is a small lexicon (with about 3000 entries) for some of the most common English words, stored in standard external form. Words not found in this lexicon are looked up in the **UDICT** lexicon [Byrd 1983, Klavans and Wacholder 1989], which has over 60,000 lemmas, in conjunction with a heuristic interface that produces entries in standard external form. The look up can use either **UDICT** in DAM form (with **UDICT** morphology) or a Prolog-based look up of the **UDICT** data stored in Prolog-readable files (together with the Slot Grammar shell morphology). When the latter method is used, the complete lexical analysis phase can be performed at a rate of about 10 milliseconds per word (or about 15 milliseconds for words coming only from the **UDICT** disk files).

---

[3] There has been a revision and improvement of the form of lexical *transfer* information is **LMT**, with more options in the overall organization and storage of the MT lexicons; but the form of source lexicons is essentially unchanged.

Work is being done to get more accurate and complete information from on-line dictionaries. Jacques Robin is working on extracting Slot Grammar-style information from Longman's. Also the work of Mary Neff on lexical data bases [Neff and Boguraev 1989] and the use of an LDB for the Collins bilingual lexicons for **LMT** [Neff and McCord 1990] may supply information useful for source analysis as well as transfer.

4 The ingredients of a Slot Grammar

In this section we discuss the types of rules (and data) that appear in the Slot Grammar **XSG** of a specific language, *i.e.* what's not in the shell. There are four major types of rules:

- A declaration of *adjunct slots* for each part of speech.
- *Slot filler rules.*
- *Slot ordering rules.*
- *Obligatory slot rules.*

In addition, there are minor types of rules (often expressed as unit clauses) dealing with language-specific aspects of ("parameters" for) phenomena treated mainly in the shell, namely extraposition, coordination, punctuation, and parse evaluation.

For most of these rule types, there is a special formalism that allows convenient reference to the parts of the phrases dealt with by the rule, and there is a rule compiler in the shell which converts the rules to Prolog clauses.

The four subsections below describe the four major rule types. Some of the minor rule types are discussed in subsequent sections.

4.1 Adjunct slots

The list of complement slots associated with a word is idiosyncratic to the word and comes from the lexical entry for the word. On the other hand, a word may have associated *adjunct* slots, and these depend only on the lexical category (part of speech) of the word. The declaration of available adjunct slots for a part of speech POS is given in the grammar by a unit clause of the form

```
category(POS,Adjuncts).
```

For example, the declaration in **ESG** for verbs is currently

```
category(verb(*),vinv.whadv.vadjp.vdet.vnoun.
                vadv.vprep.preinf.vcomment.
                vsubconj.vnfvp.vsothat).
```

(We will not explain the meanings of these adjunct slots here, although some of the names should be suggestive.)

Adjunct slots differ from complement slots in that adjuncts are always optional and in general may be filled any number of times (there may be several modifiers in a given phrase, not necessarily adjacent, which fill the same adjunct slot).

4.2 Slot filler rules

The core of a Slot Grammar is its set of slot filler rules. Each slot known to the grammar (arising from lexical entries in the case of a complement slot, or declared for a part of speech in the case of an adjunct slot) will have one or more filler rules. The basic step in parsing is to choose a slot Slot associated with (the head word of) a phrase H and to see whether an adjacent phrase M (on the left or right) can *fill* Slot by satisfying a filler rule for Slot. If so, M is added as a modifier of H, and the process is continued with the larger phrase.

A filler rule for Slot is of the form

```
Slot ==> Body.
```

The Slot on the left-hand side can be of three types: (1) a complement slot, (2) an adjunct slot, or (3) an *extraposed slot*, of the form ext(Slot0,Level), where Slot0 is a complement slot and Level is ext or norm. (The significance of Level will be explained below.) The filler rule is correspondingly called a *complement filler rule*, an *adjunct filler rule*, or an *extraposed filler rule*. The first two types of rules are called *normal* filler rules. Normal filler rules are described in the current section, and extraposed filler rules are described in Section 5. However, the form of the Body is the same for all three kinds of rules, and the rule compiler treats the body in the same way.

The Body has the same overall form as the right-hand side of a Prolog clause, *i.e.*, a combination with and's (&) and or's (|) of goal predications; but there are *special goals* that can refer to parts of the filler phrase or the head phrase. Special goals, known to the filler rule compiler, are compiled in a special way, and all other goals are compiled as themselves.

As an example, a filler rule for the object and indirect object slots might be

```
obj ==> f(noun(*,acc.*,*)).

iobj ==> f(noun(*,dat.*,*)).
iobj ==> f(prep(to,*,*)).
```

Here f(Feas) is a special goal that requires that the filler phrase has feature structure Feas. As another example, a rule for the subject slot

```
subj ==> f(noun(*,nom.Num,*)) &
         hf(verb(fin(*,Num,*,*))).
```

can require subject-verb agreement in number. Here the special goal hf(Feas) requires that the *head* phrase (or *higher* phrase) has feature structure Feas.

Special goals are two types: *selector goals* and *feature-changing goals*. Selector goals arise from the following scheme. There are several *selector* predicates for accessing parts of a phrase data structure. For example, f(P,Feas) says that phrase P has feature structure Feas. This selector is of course defined simply by the unit clause:

```
f(phrase(*,*,*,Feas,*,*,*),Feas).
```

A more complex example is eslot(P,Slot,X), which says that phrase P has an empty (unfilled) complement slot Slot with marker X. This is defined by

```
eslot(phrase(*,*,*,*,Frame,*,*),Slot,X) ←
    member(slot(Slot,*,X),Frame) & var(X).
```

(This can be used both for generating empty slots and for testing whether a given slot is empty.)

For all of these selector predicates, the first argument is a phrase. In filler rules, the filler phrase and head phrase are *implicit*. So the simple scheme is that for each selector predicate

```
pred(P,X1,...,Xn)
```

where P is the phrase argument, we get two special goals

```
pred(X1,...,Xn)
hpred(X1,...,Xn)
```

where the phrase argument is *understood* to be the filler phrase in the first case and the head phrase in the second case. In this way, the special goals f(Feas) and hf(Feas) arise from the selector predicate f(P,Feas).

Selector predicates can also be used in their full forms in filler rules. One of the modifiers of the filler or head phrase could be selected (by use of a special goal), and then a full-form selector could be applied explicitly to this modifier phrase. However, it is not common to have to do this.

Feature-changing goals will not be dealt with in the current paper; for details, see [McCord 1989b]. Also, see this reference for a description of the compilation of filler rules.

4.3 Slot ordering rules

There are two kinds of slot ordering rules: (1) *head/slot ordering rules*, expressing ordering of slots with respect to the head word, and (2) *slot/slot ordering rules*, expressing ordering of slots with respect to other slots. Although these rules are stated in terms of *slots*, the interpretation is of course in terms of *fillers* of slots and their left-to-right surface order. Every time a normal (non-extraposed) slot filler rule is applied, the parser checks that the position of the new filler is allowed by the ordering rules.

In some cases, ordering of slots is unconditional, but in other cases it can depend on characteristics of filler phrases or the head phrase. For example, in English the subj slot can be on the right of the head verb in yes/no questions when the head verb is an auxiliary (auxiliaries are treated as higher verbs in **ESG**).

Head/slot ordering rules are of either of the forms:

```
lslot(Slot) ← Body.

rslot(Slot) ← Body.
```

The antecedent side Body (and the arrow) may be omitted. These rules say respectively that Slot is a *left slot* (or *right slot*), i.e., that its filler is a left (or right) modifier of the head word, under the conditions of Body. The Body has the same form as the body of a filler rule, except that feature-changing goals are not allowed. Thus, selector goals can refer to any parts of the filler phrase or the head phrase.

As an example, constraints on the subj slot in English can be expressed by the rules

```
lslot(subj) ← hf(verb(fin(*,*,*,*))).

rslot(subj) ← hf(verb(fin(*,*,*,ind:q:*))) &
              hsense(Head) & finaux(Head).

finaux(be).
finaux(have_perf).
finaux(will).
   ...
```

The first rule says that subj can be a left slot for a finite verb, and the second rule says that subj can be a right slot in a question sentence (feature q) if the verb is a finite auxiliary. (Absence of the feature q signals a declarative sentence.)

A slot/slot ordering rule is of the form:

```
LSlot<<RSlot ← Body
```

where the Body may be omitted. This means that every filler for LSlot must precede every *different* filler for RSlot under the conditions of Body. The point in saying "different" here is that LSlot can equal RSlot, and then the ordering constraint says that LSlot can have no more than one filler. (Two different fillers would have to precede each other.) For example, we could have the following rule for the noun determiner (adjunct) slot ndet:

```
ndet << All.
```

Here All is just a Prolog variable, so this means that ndet precedes every slot, including itself. Therefore, there is at most one determiner, and this must precede every other noun modifier.

The body of a slot/slot ordering rule can contain selector goals allowing reference to the phrases involved, namely the filler of LSlot, the filler of RSlot, and the head phrase. Correspondingly, the selector goals are of the form lpred, rpred, and hpred, where pred is a selector predicate. For example, lf(Feas) selects the feature structure of the filler of LSlot, and hframe(Frame) selects the slot frame of the head phrase.

Examples of conditional slot/slot ordering rules are the following ones, expressing relative ordering of the direct and indirect objects of verbs:

```
iobj<<obj ← lf(noun(*,*,*)).
obj<<iobj ← rf(prep(*,*,*)).
```

These say that the indirect object precedes or follows the direct object according as the indirect object is a noun phrase or a prepositional phrase. (Cases of heavy direct objects that can follow a PP indirect object are ignored here.)

4.4 Obligatory slot rules

A complement slot may be *optional* or *obligatory*. (Adjuncts are always considered optional.) For a phrase to become a modifier of another phrase, all of the obligatory slots of the modifier phrase (after possible extraposition of one of its slots, as described in Section 5) must be filled. Also, an allowable top-level analysis must have all of its obligatory slots filled. An optional slot need not get a filler. As an example, the verb *eat* might have an optional object slot, but the verb *put* might have an obligatory object slot.

There can be optional and obligatory versions of the same slot. In fact, in internal form a slot slot(Slot,Ob,X) is obligatory when Ob=ob, and optional when Ob=op. There is a convention (known by the lexical compiler) for slots shown in the external lexicon, that an obligatory form of a slot is obtained from the ordinary name by adding the character 1. Thus obj1 is the obligatory form of obj, and its internal form is slot(obj,ob,X). In the grammar, one just refers to obj.

Thus slots in particular slot frames of words may be obligatory because the words' lexical entries specify this, but it is also possible to write general rules in the grammar that require slots to be obligatory, either absolutely or conditionally. An obligatory slot rule is of one of the forms:

```
obl(Slot) ← Body.

obl(Slot,Slot1) ← Body.
```

As before, the **Body** and arrow may be omitted. The first type of rule rule says that **slot** is obligatory under the conditions of **Body**. The second type of rule, a *relative* rule, says that **slot** is obligatory *if* **slot1** is filled and the conditions of **Body** hold. The body has a form similar to that of the body in our previous rules, allowing reference to relevant phrase characteristics through selector goals. In the first type of rule, one may refer to the head phrase through goals of the form **hpred**. Such goals are allowed in the second type of rule, as well as goals (of the form **pred**) which refer to the filler of **slot1**.

Examples of simple obligatory slot rules for English are the following:

```
obl(subj) ← hf(verb(fin(*,*,*,*))).

obl(obj,iobj) ← f(noun(*,*,*)).

obl(objprep).
```

The first rule says that the subject is obligatory in a finite verb phrase. The second rule says that the direct object is obligatory if the indirect object is filled by a noun phrase. The third rule says that the object of a preposition is obligatory unconditionally.[4]

5 Extraposition

Recall from Section 2 that in a phrase structure

```
phrase(X,H,Sense,Features,SlotFrame,Ext,Mods),
```

the argument **Ext** is used to hold *extraposed* slots, *i.e.*, slots that can be filled by left-extraposed phrases like *who* in *Who did Alice try to find*. The list **Ext** consists of slots in internal form, and it can contain more than one element, as we will see below.

The shell takes care of several of the problems with extraposition, but there are two ingredients in the grammar itself that deal with it:

1. a declaration of *extraposer* slots, and

2. extraposed filler rules.

Extraposer slots are slots that allow extraposition *out of* their fillers. Consider the above example

---

[4] As indicated above, obligatory slots need not be filled immediately on the level of their head words, but may be extraposed and filled on a higher level. Thus sentences like *Which chair was John sitting in?* are allowed even though **objprep** is obligatory.

Who did Alice try to find?

Here the verb phrase *to find* fills a slot `obj(inf)` for *try*. The `obj(inf)` slot is declared to be an extraposer by a clause

```
extraposer(obj(inf)).
```

in the grammar. This tells the parser that the unfilled `obj` slot for *find* can be extraposed to the level of *try*, becoming the sole member of the Ext argument of the `phrase` structure of *try*. It also remains in the frame of *find*. (The restrictions on what slots (like `obj`) can be extraposed will be described below.) Next, the phrase *try to find* fills a slot `auxcmp(inf(bare))` for *did*. This slot is also declared an extraposer in the grammar. The still-unfilled `obj` slot is extraposed still higher, to the level of *did*. Then finally an extraposed filler rule allows *who* to fill this slot. The link of *who* to the object of *find* is shown in the `phrase` structure, because the marker of the `obj` slot down in the frame for *find* is unified with the marker for *who*.

The slots that can always be extraposed through a slot declared as an **extraposer** are the slots that are basically objects: `obj`, `pobj(Prep)`, and `objprep` (the third one is the object of a preposition). There are others, but we will not list them all.

What about the extraposition of subjects? We need to capture the fact that we can say

Who did you say found the book?

but not

Who did you say that found the book?

If a slot s is declared a **strongextraposer**, then the `subj` slot (as well as the object-type slots named above) can be extraposed through s. In addition, the `subj` slot can be extraposed through an extraposer *if it has already been extraposed* (is in the Ext argument as opposed to the Frame argument of the filler phrase).

So let us see how this allows us to handle the preceding two examples. In the sentence, *Who did you say found the book*, the verb phrase *found the book* fills a slot `obj(fin)` of *say*, and there is a declaration

```
strongextraposer(obj(fin)).
```

Hence the unfilled **subj** slot of *found* can be extraposed through `obj(fin)` to the level of *say*. When the verb phrase *say found the book* fills the `auxcmp(inf(bare))` slot (an extraposer) of *did*, the **subj** slot of *found* can be extraposed, since it has already been extraposed. Then it is available to be filled by *who* using an extraposed filler rule.

Before looking at the second, ungrammatical example above, let us look at a similar example that is grammatical:

What did you say that Alice found?

Here *that* is viewed as a special type of subordinating conjunction with the feature structure fsubconj. It has a single complement slot, fvcomp, which can be filled by a finite verb phrase. Furthermore, fvcomp is declared an extraposer, so that when the verb phrase *Alice found* fills fvcomp, the obj slot can be extraposed to the level of *that*. Now the obj(fin) slot of *say* has a filler rule allowing it to be filled by a phrase with feature structure fsubconj, so that it can be filled by *that Alice found*, with further extraposition of obj, since obj(fin) is a (strong) extraposer.

Now let us return to the example

Who did you say that found the book?

The point is simply that the complement slot fvcomp of *that* is not declared a *strong* extraposer. Hence there is no chance for the subj slot of *found* to be extraposed. In fact, the parser does not allow *found the book* to fill the fvcomp slot, because there is a slot (its subj) which (1) cannot be extraposed and (2) is obligatory. (The subject slot is declared obligatory for finite verb phrases.)

It should be noted that the only types of slots that can be extraposed are complement slots; adjunct slots are not extraposed. If one wanted to handle extraposed attachments of wh-adverbials as in *When did you invite him to speak?* (where one reading links *when* to *speak*), then one should allow extraposition of the markers of phrases. Making the marker of an embedded verb phrase available on a higher level would allow extraposed attachment of wh-adverbials to the embedded phrase. Currently, however, the system does not do this; adverbials are linked only to their surface head verbs.

The system does allow extraposition of prepositional object slots (objprep) through adjunct slots, allowing correct parsing of examples like

Who did you say John was sitting with?

where *who* fills the extraposed objprep of *with*, and the *with*-phrase fills an adjunct slot for *sitting*.

It was mentioned above that the extraposed slot list Ext can contain more than one element. This occurs with an ambiguous sentence like

Which horse do you want to win?

Here we give the verb *want* the slot frame `subj.obj.comp(inf)` (with `obj` optional), where the VP *to win* fills `comp(inf)`. The ambiguity results from the fact that *which horse* can fill either the `obj` slot of *want* or the `obj` slot of *win*. (As described below in Section 7 on *Implicit subjects*, the system decides that the implicit subject of *win* is *which horse* in the first case, and *you* in the second case.) The extraposed slot list of *do* contains two slots because both the object of *win* and the object of *want* are extraposed to the level of *do*. Thus *which horse* can fill either of these two slots, and **ESG** produces two parses.

As indicated in Section 4.2, the filling of extraposed slots is handled by *extraposed filler rules*, which are of the form

```
ext(Slot,Level) ==> Body.
```

The argument `Level` is `ext` if the slot is chosen from the extraposed slot list `Ext` of the head phrase, and it is `norm` if the slot is chosen from the slot frame of the head phrase. We need to allow the latter for examples like the filling of the relative pronoun *that* in *the man that John saw*.

In **ESG** there are just two extraposed filler rules. One of them fills object-type slots and `subj` by *wh*-noun phrases (including relative pronouns). The other fills the `pobj(Prep)` slot by *wh*-prepositional phrases. In both cases, the marker of the *wh*-element is saved in the feature structure of the head verb phrase, so that when this verb phrase acts as a relative clause (filling the adjunct slot `nrel` for nouns), the *wh*-marker can be linked to the marker of the head noun.

What about *abbreviated* relative clauses, as in *the book I was trying to find*? In such cases, extraposition of slots proceeds just as for full relative clauses, so that in the example the `obj` slot of *find* is extraposed to the level of *was*. But there is no overt relative pronoun to trigger filling of this slot. Instead, the filling is triggered in the following way. There is a special adjunct slot `abbnrel` for nouns which is filled by abbreviated relative clauses. The filler rule for `abbnrel` looks for an unfilled normal or extraposed slot in the filler verb phrase that *could* receive a relative pronoun, and then "fills" this slot in the sense that its marker is unified (essentially) with the marker of the head noun. In addition, `abbnrel` employs some constraints on the form of the proposed abbreviated relative clause, which are useful in cutting down on spurious partial analyses in parsing.

6 Coordination

The following method for handling coordination was outlined in [McCord 1980], and was implemented (with improvements) in the recent adaptation of Slot Grammar to logic programming.

First let us discuss the types of coordinated phrases analyzed by the system, and the form of phrase analyses produced for them. The main form for such a phrase is

```
LM  Preconj  LC  Conj  RC  RM
```

where the substrings indicated are as follows. Conj is a coordinating conjunction (like *and* or *or*), or a punctuation symbol (like a comma) used in the capacity of a coordinating conjunction. Preconj is an optional *preconjunction* that can accompany Conj (like *both* for *and*). LC and RC are the *left* and *right conjuncts*, respectively. Each of these conjuncts consists of a single phrase, although it may not be *satisfied* (it may have unfilled obligatory slots). LM and RM are the (optional) *left* and *right common modifiers*, respectively (each of these may be represented by several phrases). Some examples:

```
The man   sees   and   probably hears   the car.
-------   ----   ---   --------------   -------
  LM       LC    Conj       RC            RM


John sees   and   Mary hears   the car.
---------   ---   ----------   -------
   LC       Conj      RC          RM
```

The specification of coordinating conjunctions and their associated preconjunctions is of course language-specific, and is given in the grammar[5] by clauses like:

```
conj(and).
```

```
preconj(and,both,bothand).
```

We separate preconj from conj because in some languages a given conjunction can have more than one preconjunction.

The phrase structure produced by the system for a coordinated phrase LM Preconj LC Conj RC RM is as follows.

(1) The head (sense) is basically the conjunction Conj, but is actually the following compound term:

```
coord(Conj1,LHead,RHead).
```

If a preconjunction Preconj is present, then Conj1 is defined by

---

[5]   This information is put in the grammar instead of the lexicon because of the special and limited nature of these words.

```
preconj(Conj,Preconj,Conj1).
```

Otherwise conj1=conj. The second and third arguments of coord are the heads of the conjuncts LC and RC.

(2) The feature structure Feas of the coordinated phrase is manufactured from the feature structures LFeas and RFeas of the conjuncts by a procedure

```
coordfeas(Conj,LFeas,RFeas,Feas)
```

specified in the grammar. (This constitutes the main language-specific information regarding coordination.) Typically, LFeas and RFeas will be required to be rather similar, and usually to be of the same part of speech, with the result Feas sharing at least this part of speech. For example, in **ESG** a finite verb phrase will coordinate only with another finite verb phrase (although their tenses, etc., are allowed to differ), and the result is a finite verb phrase. Thus, so far as features go, coordinating conjunctions are "parasitic" on their conjuncts.

(3) The slot frame of the coordinated phrase is manufactured by a procedure coordframe in the parser, using both the slot frames and the extraposed slots of the conjunct phrases. The idea is to "factor out" common slots in the conjuncts, raising them up to the level of the new coordinated phrase, making them available for filling on this level. In particular, common modifiers (in LM and RM) may fill slots in this factored-out frame.[6] In the last example sentence above, *the car* fills a slot obj factored out from the frames of *sees* and *hears*. The procedure coordframe is the core of the treatment of coordination in the parser, and will be described in more detail below.

(4) The extraposed slot list Ext of the coordinated phrase is empty. Extraposed slot filling *can* be done on this level (or higher), but it is just done using slots stored in the frame. (There is no need to keep extraposed slots separate from normal slots at this point, because the "real" frame positions of slots are stored below in non-conjoined phrases.)

(5) The left modifiers of the coordinated phrase are as follows. The rightmost one is of the form lconjunct:LP where LP is the phrase representing the left conjunct LC. The initial left modifiers (if any) represent the common left modifiers LM. The form of the right modifiers is symmetric, with RC filling the slot rconjunct.

Now let us look in more detail at the work of coordframe. The first thing to note is that in the process of factoring out slots, we must be ready to consider (unfilled) slots LSlot and RSlot of the conjuncts to produce a common slot even when LSlot and RSlot are not exactly the same. For instance, the slot obj(fin) mentioned in the preceding section can be filled not only by finite verb phrases, but also by noun phrases (it is a complement slot of *believe*, for example). This could be considered compatible with an obj slot, for the purposes of factoring out, and the result would be the "greatest common denominator" slot obj. Therefore we use a procedure

---

[6] Such common modifiers will of course be complement modifiers, but common modifiers could also fill adjunct slots selected for the coordinated phrase solely on the basis of its feature structure.

```
coordslot(LSlot,RSlot,Slot)
```

which can produce a "g.c.d" slot Slot from LSlot and RSlot.

The next observation is that the results of extraposition are of direct value for coordination; extraposed slots can be used in the process of factoring out, becoming filled (in common slot form) by a phrase that is not fronted. Consider an example of Woods [1973]:

John drove his car through and completely demolished a plate glass window.

ESG handles this as follows. In the left conjunct *drove his car through*, the objprep slot of *through* is extraposed to the level of *drove*. This slot can be paired by coordslot with the obj slot in the right conjunct *completely demolished*, producing a common slot obj, which gets filled by *a plate glass window*. The subj slots (non-extraposed) of the two conjuncts are factored out and filled by *John*. The parse produced by ESG is shown in Figure 2.

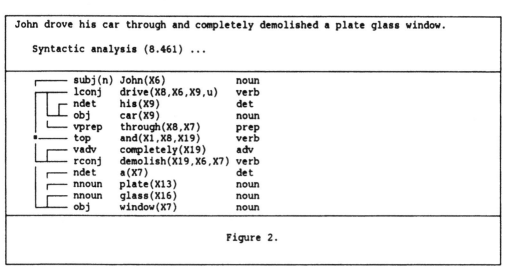

```
John drove his car through and completely demolished a plate glass window.

   Syntactic analysis (8.461) ...


────── subj(n)  John(X6)            noun
────── lconj    drive(X8,X6,X9,u)   verb
────── ndet     his(X9)             det
────── obj      car(X9)             noun
────── vprep    through(X8,X7)      prep
────── top      and(X1,X8,X19)      verb
────── vadv     completely(X19)     adv
────── rconj    demolish(X19,X6,X7) verb
────── ndet     a(X7)               det
────── nnoun    plate(X13)          noun
────── nnoun    glass(X16)          noun
────── obj      window(X7)          noun
```

Figure 2.

Note that the marker X6 for John is also the subject variable for drive and for demolish, and the marker X7 for window is the object variable for through and for demolish.

One can make up examples involving several levels of extraposition, which are handled in a similar way by ESG.

Because of this double use of extraposition, the extraposed slot lists of the two conjuncts are pooled together with (concatenated onto) their normal slot frames for the work of coordframe. Let us call these combined frames for the left and right conjuncts LFrame and RFrame, respectively.

Then coordframe succeeds in producing a factored-out frame Frame from LFrame and RFrame, under the following conditions:

1. Whenever an unfilled slot LSlot of LFrame can be paired by coordslot with an unfilled slot RSlot of RFrame, their markers are unified, and their Ob components are suitably combined. The resulting (factored out) slot is made a member of Frame.

2. Any unfilled slot of LFrame or RFrame that is not paired as in (1) must be optional (where we consider verb subjects obligatory).

The coordination system handles a bit more than the pattern

    LM Preconj LC Conj RC RM

specified above, namely the following three variations.

(a) If Conj is a lexical coordinating conjunction (not punctuation), then it may be preceded by a comma or semicolon.

(b) An adverb is allowed between Conj and RC (and it is made the leftmost right modifier of the coordinated phrase). Such adverbs are allowed in coordination of phrases of any category, even ones not normally modified by adverbs. (This is due of course to the fact that coordination of any category can abbreviate sentential coordination.)

(c) The construction

    Conj RC

(where Conj is lexical) is considered a valid phrase by itself. This is useful not only because such phrases do occur, but also because in full coordinated phrases the Conj RC portion can be enclosed in *brackets* (parentheses and other paired symbols), as in

   The director will decide on (or soon consider) the case.

There are general rules in the system for dealing with bracketing, and these apply to phrases, so that examples like the preceding are handled if Conj RC is made a phrase. Therefore, even for building a full coordinated phrase, a preliminary step in the parser is to make the Conj RC portion a phrase, giving it a chance to be bracketed. Then this structure is used as input to the building of the (more symmetric) structure for the full coordinated phrase described above.

7 Implicit subjects

It is useful to identify implicit subjects of non-finite verb phrases when possible. This is done (in some cases) by the Slot Grammar shell for non-finite verb phrase complements of verbs, in a pass performed after complete phrase analyses of the input string have been obtained. The method used does not always give correct results, but seems to be a useful approximation.[7]

Suppose we are given a verb phrase VP1, and a non-finite verb phrase complement VP2 of VP1, where the grammatical subject slot S2 of VP2 is unfilled. We "identify" S2 by unifying its marker with the marker of a suitably chosen slot S1 in the frame of VP1. In many cases, S1 will already be filled, so that S2 will be related then to an overt filler.

So the problem is in choosing the higher slot S1. The full statement of the rule we follow is given below, but the basic idea (to which there are exceptions) is the following. If an *object-type* slot is present and filled in the frame of VP1, then S1 is taken to be this slot; otherwise S1 is the subj slot of VP1. The object-type slots are obj, pobj(Prep), and iobj.

As an example, suppose that *want* has the slot frame

    subj.obj.comp(inf)

(where obj is an optional slot). Then in *John wants to leave*, the subject of *leave* would be linked to the subject of *wants* (because obj is not filled). But in *John wants Bill to leave*, it is linked to the obj (*Bill*) of *wants*.

There are exceptions to the basic rule stated above, due just to the nature of the head verb of VP1.[8] In both of the sentences *John promised to leave* and *John promised Bill to leave*, the lower subject is linked to the higher subject. Such verbs are specially marked ss ("subject-to-subject") in the lexicon. Some verbs, like *advise*, refuse to accept a link to their subjects (in active form), and these are specially marked so ("subject-to-object") in the lexicon. In general, verbs (like *expect*) may have neither of the features ss, so.

The complete statement of our implicit subject rule is as follows. (Assume the same meanings for VP1, VP2, and S2 as above.)

1. If VP1 is passive (marked pastpart), then S2 is linked to the grammatical subject of VP1;
2. otherwise, if an object-type slot S1 is present and filled in the frame of VP1 and the head of VP1 is not marked ss, then S2 is linked to S1;
3. otherwise, if the head of VP1 is not marked so, then S2 is linked to the subject of VP1;
4. otherwise, no link is made.

There are still difficulties with this rule, because the correct linking can also depend on the nature of the embedded verb phrase, and on contextual conditions. So we have only an approximation.

---

[7]   A similar method was used in the logic grammar described in [McCord 1982].
[8]   I wish to thank Hubert Lehmann and Brigitte Barnett for a useful conversation in this regard.

To see why implicit subject identification is done *after* the final parse is obtained, consider the example discussed in Section 5 above: *Which horse do you want to win?* The point is simply that we do not know whether the object slot of *want* is filled until we have the complete parse, where we have decided whether *which horse* fills this slot or the object slot of *win*.

An alternative approach (followed in an earlier version of the system) allows the determination of implicit subjects *during* parsing, but at the expense of using more word frames for verbs like *want*.

8 The basic parsing algorithm

The overall idea of the parser is the following. After lexical analysis of the input word string, resulting in *word analyses* (as described in Section 3), the words are processed left-to-right, building up **phrase** analyses of substrings. All satisfied **phrase** analyses of the entire input string are considered valid parses. (A phrase is satisfied if all its obligatory slots are filled.)

More specifically: In the left-to-right processing, when a new word (or multiword) is encountered, we look at each word frame for it and construct the corresponding initial **phrase** having no modifiers and an empty extraposition (**Ext**) argument. This phrase is considered a new *result*. A *result* records not only a **phrase** structure, but also the left and right boundaries of the phrase in the word string, a term representing a parse evaluation (to be described in Section 9) and a *state*, representing what kind of modifiers (left, right, or extraposed) the phrase has received. Specifically, **state** will be 0 if the phrase has no right or extraposed modifiers, but may have left modifiers; **state** will be 1 if the phrase has some right modifiers, but no extraposed modifiers; and **state** will be 2 if the phrase has some extraposed modifiers. Of course **state** is set to 0 when an initial result is created from a word frame.

Whenever a new result R is obtained, it is *inserted* into the store of previous results by a procedure **insert**. Basically, this means the following. Given R, **insert** selects a previous result L whose right boundary coincides with the left boundary of R, and tries to *combine* L and R. *Combining* will be explained in the next paragraph. If the combination succeeds, then a new *result* is obtained, and this is recursively *inserted*. Backtracking in the actions just described is forced by a **fail** goal in the definition of **insert**. Thus every adjacent result L will be tried, and all ways of combining L and R will be tried.

Two adjacent results can be *combined* in four different ways: (1) The phrase on the left can become a *normal slot filler* of the one on the right; (2) *vice versa*; (3) the phrase on the left can become an *extraposed slot filler* of the one on the right; or (4) the two phrases can be combined in coordination.

Coordination has been discussed above, so by now our exposition of the basic parsing algorithm is reduced to an explanation of the procedure for slot filling. The procedure involved is

```
fillslot(Side,Level,MPhrase,HPhrase,Phrase)
```

(described in slightly simplified form). This lets MPhrase fill a slot in HPhrase on side Side, with result Phrase. The slot filling is normal or extraposed according as Level is norm or ext.

There are three rules for fillslot, corresponding to the three types of slot filler rules invoked: (a) normal filler rules for complement slots, (b) (normal) filler rules for adjunct slots, and (c) extraposed filler rules (for complement slots).

For each of the three fillslot rules, the basic steps involved are the following:

1. Choose an unfilled slot of the appropriate type.

2. Apply the appropriate filler rule.

3. Check ordering constraints, for normal slot filling (not done for extraposed filling).

4. Apply extraposition from the modifier phrase if possible (not done for extraposed filling).

5. Check that the modifier phrase is satisfied (modulo slots extraposed from it).

6. Bind marker variables.

Of course the new phrase resulting from the modification is obtained by "updating" the head phrase with the new modifier, possible new extraposed slots, and a possible change in its feature structure.

Most of the operation of the six steps should be clear from the discussion in the preceding sections, but the following comments are worthwhile.

The first of the six steps adds non-determinism to fillslot. (In general, there can be more than one unfilled slot.)

In the actual code, steps 4 and 5 are combined, because it is more efficient to run through the modifier's slots only once. For each unfilled slot Slot in the modifier's slot frame or extraposition list, if Slot can be extraposed (Section 5) then this is done. Otherwise Slot must not be obligatory in the modifier (Section 4.4).

Step 3 (checking ordering constraints) is performed as follows. Call the head phrase P, and suppose that the slot being filled by fillslot is Slot, and its modifier MP is on the *right* of P. (The treatment of a left modifier is essentially symmetric.) Then the following two checks are made, using the slot ordering rules of the grammar (in their compiled forms). (a) Slot must be a right slot for MP modifying P (*i.e.* an rslot rule for these items must hold). (b) It must *not* be the case that there is a filled slot Slot1 such that the slot/slot ordering relation Slot << Slot1 holds. (Since the filler MP of Slot is going to be placed on the right, this would be a conflict.) There is a slight complexity in the search for a *filled* slot Slot1 in the second check. The associated modifier could be simply among the immediate modifiers of P, but P could be coordinated. If so, one must also look down in the conjuncts.

9 Parse evaluation and pruning

Our parse evaluator is based on a partial order

```
R1 betterthan R2
```

defined on the parse space (the set of all results R). During parsing, if a new result R is obtained and there is a previous result that is betterthan R, then R will not be considered further (will not be inserted). If R is not ruled out in this way, then any previous results that are worse than R will be deleted.

In addition, an equivalence relation is defined on the parse space, expressing broad similarity of feature structure in results. Results can be related by betterthan only when they are in the same equivalence class, so that parse pruning is done independently in each equivalence class. Within an equivalence class, betterthan is a total order, based on numerical scoring.

Currently, betterthan expresses preferences in three different dimensions:

1. close attachment,

2. preference for complements (over adjuncts), and

3. parallelism in coordination.

The component of the numerical scoring function that controls for close attachment is Heidorn's [1982] parse metric. These three aspects of preference *can* be at odds with one another. For example, in the sentence *John sent a file to Bill*, close attachment would favor attachment of *to Bill* to *file*, but complement preference would favor attachment to *sent* (because of the iobj slot for *sent*). To settle this conflict, betterthan is defined so that complement preference completely dominates close attachment (*i.e.*, close attachment is considered only if there is no decision by complement preference).[9]

The relation betterthan for results is defined in terms of the *evaluation structures* stored in results. We define these structures now, and a partial order, also called betterthan, on them. Then a result R1 is taken to be betterthan a result R2 if and only if R1 and R2 have the same boundaries and the evaluation structure of R1 is betterthan that of R2.

An *evaluation structure* for a result is a term of the form

```
eval(H,F,Score)
```

---

[9] The combined treatment of complement preference and close attachment given here has some similarities (in results, but not in the algorithm) with the treatment in [Wilks, Huang and Fass 1985]. The ideas were also implemented in ModL [McCord 1986, 1989a] in still another way. The lexical information necessary for setting up complement preferences in [Wilks, Huang and Fass 1985] was represented in terms of semantic cases, but it can also be represented in terms of syntactic cases (our slots) with associated semantic type requirements. Semantic type checking is not currently done systematically in the Slot Grammar system (although it could be added). Nevertheless, the purely syntactic preference methods described in this section often produce parses suitable for use in MT.

where the three components are as follows.[10] (1) H is the word number of the head word of the result (phrase). (2) F, the *basic feature* of the phrase, is obtained from the feature structure **Feas** of the phrase as follows: In most cases, F is the principal functor of **Feas** (*i.e.*, the part of speech), but if **Feas** is **verb(Infl)** then F is the principal functor of **Infl**. (3) **Score** is a real number giving the basic score of the evaluation.

Before describing how the score is computed, let us say immediately that **betterthan** is defined on evaluation structures by:

```
eval(H1,F1,S1) betterthan eval(H2,F2,S2)
    iff H1=H2 and F1=F2 and S1<S2.
```

Thus two evaluation structures (and hence two results with the same boundaries) are comparable by **betterthan** if and only if they have the *same* heads and basic features. And within the equivalence class of results having the same heads, basic features and boundaries, we get a *total* ordering defined by the score.

Now let us turn to the computation of the score of a result (the score component of its evaluation structure). This is defined by a formula

```
Score = CPEval + AdEval + HEval.
```

The purposes of the three components are as follows.

- **CPEval** controls for parallelism in coordinated phrases.

- **AdEval** controls for complement preference, and is *roughly* the total number of adjuncts in the phrase.

- **HEval** is the *Heidorn evaluation* of the phrase [Heidorn 1982], which controls for close attachment.

Let us look at the definitions of the three components. (Recall that a lower score is a better score.)

The Heidorn evaluation H(P) of a phrase P is defined recursively as the sum of all terms 0.1*(H(Mod)+1), where **Mod** varies over the modifiers of P. (One need not state the base of this recursive formula separately, since one arrives eventually at phrases with no modifiers, and then the sum over the empty list is understood to be zero.) Actually, the factor 0.1 used in the recursive formula is just a default. An exception rule, associated with a slot in the grammar, can specify a different factor. Also, exception rules can change the basic formula applied, in a way that will not be described here.

As mentioned above, the **AdEval** component of a score (used for complement preference) is defined roughly as the total number of adjunct modifiers in the phrase (on all levels). Thus we could say that roughly for each adjunct slot filling, 1 is added to the score, and for each complement slot filling, 0 is added to the score (in this connection). This is the *default*, but it

---

[10]   The exposition here is slightly simplified over what is in the actual system.

can be overridden for a slot slot if there is an *evaluation rule* for slot in the grammar. Such a rule looks like a Prolog clause for the predicate

```
adeval(Slot,A)
```

where A is a number to be added for the slot in place of the default. The body of such a clause (if present) can contain selector goals (just as in slot filler rules) which refer to any parts of the filler phrase or the higher phrase.

Now let us look at the CPEval component of a score. When two phrases are coordinated, coordfeas requires that their feature structures be similar (normally having the same part of speech). But it may not require that the feature structures be *exactly* the same. For instance, the feature structures of prepositional phrases exhibit the head preposition, and we do not want to *require* that coordinated PPs have the same head preposition. But when we are deciding which of several PPs to coordinate a given one with, it is *better* (for greater parallelism) to choose one with the same head. This happens if the feature structures of the PPs are exactly the same. To capture this preference in the score for a coordinated phrase we add 0 or 1 to the CPEval component according as the feature structures are "exactly the same" or not.[11]

The CPEval component controls for another aspect of parallelism in coordination, namely parallelism in the configurations of modifiers of each conjunct. The idea, as implemented currently, is just that it is preferrable that the conjuncts have nearly the same number of left modifiers and nearly the same number of right modifiers. Specifically, for each coordination, the number

```
|#LMods1-#LMods2| + |#RMods1-#RMods2|
```

is added to the CPEval component of the score. Here #LMods1 and #LMods2 are the numbers of left modifiers in the left and right conjuncts, respectively, and #RMods1 and #RMods2 are the numbers of right modifiers in the left and right conjuncts, respectively. (There is a slight amendment to this if one of the conjuncts itself is coordinated.) As an example, for the phrase *the old men and women*, the analysis [the old] [men and women] gets 0 added to CPEval, but the analysis [the old men] and [women] gets 2 added to CPEval.

Finally, the CPEval component controls for similarity in the slot frames of the two conjuncts. Recall from Section 6 that the procedure coordframe does require some similarity in the slot frames of the two conjuncts, but the frames may differ in the presence of some *optional* slots. However, such slots can create a penalty that is added to the CPEval component. There are defaults for such penalties, and one may also specify exceptions in the grammar.

---

[11] This notion can be defined recursively as follows. (1) Any atomic term is exactly the same as itself. (2) Any two variables are exactly the same (they need not be unified). (3) Two compound terms are exactly the same if they have the same principal functor and the same arity, and their corresponding arguments are exactly the same. There is a built-in predicate =*= in VM/Prolog that checks this directly.

References

Bernth, A. [1989] "Discourse Understanding In Logic," *Proc. North American Conference on Logic Programming*, pp. 755-771, MIT Press.

Bernth, A. [1990] "Treatment of Anaphoric Problems in Referentially Opaque Contexts," these Proceedings.

Byrd, R. J. [1983] "Word Formation in Natural Language Processing Systems," *Proceedings of IJCAI-VIII*, pp. 704-706.

Colmerauer, A. [1978] "Metamorphosis Grammars," in L. Bolc (Ed.), *Natural Language Communication with Computers*, Springer-Verlag.

Heidorn, G. E. [1982] "Experience with an Easily Computed Metric for Ranking Alternative Parses," *Proceedings of Annual ACL Meeting, 1982*, pp. 82-84.

Jensen, K. and Heidorn, G. E. [1982] "The Fitted Parse: 100% Parsing Capability in a Syntactic Grammar of English," Research Report RC 9729, IBM Research Division, Yorktown Heights, NY 10598.

Klavans, J. L. and Wacholder, N. [1989] "Documentation of Features and Attributes in UDICT," Research Report RC14251, IBM T.J. Watson Research Center, Yorktown Heights, N.Y.

Lappin, S. and McCord, M. C. [1989] "A Syntactic Filter on Pronominal Anaphora for Slot Grammar," to appear in the Proceedings of ACL 90.

McCord, M. C. [1980] "Slot Grammars," *Computational Linguistics*, vol. 6, pp. 31-43.

McCord, M. C. [1982] "Using Slots and Modifiers in Logic Grammars for Natural Language," *Artificial Intelligence*, vol. 18, pp. 327-367.

McCord, M. C. [1985] "Modular Logic Grammars," *Proc. 23rd Annual Meeting of the Association for Computational Linguistics*, pp. 104-117, Chicago.

McCord, M. C. [1986] "Design of a Prolog-based Machine Translation System," *Proc. of the Third International Logic Programming Conference*, pp. 350-374, Springer-Verlag, Berlin.

McCord, M. C. [1987] "Natural Language Processing in Prolog," in Walker et al. [1987].

McCord, M. C. [1988] "A Multi-Target Machine Translation System," *Proceedings of the International Conference on Fifth Generation Computer Systems 1988*, pp. 1141-1149, Institute for New Generation Computer Technology, Tokyo, Japan.

McCord, M. C. [1989a] "Design of **LMT**: A Prolog-based Machine Translation System," *Computational Linguistics*, vol. 15, pp. 33-52.

McCord, M. C. [1989b] "A New Version of Slot Grammar," Research Report RC 14506, IBM Research Division, Yorktown Heights, NY 10598.

McCord, M. C. [1989c] "A New Version of the Machine Translation System **LMT**," *J. Literary and Linguistic Computing*, vol. 4, pp. 218-229.

McCord, M. C. [1989d] **"LMT,"** *Proceedings of MT Summit II*, pp. 94-99, Deutsche Gesellschaft für Dokumentation, Frankfurt.

McCord, M. C. and Wolff, S. [1988] "The Lexicon and Morphology for LMT, a Prolog-based MT system," Research Report RC 13403, IBM Research Division, Yorktown Heights, NY 10598.

Neff, M. S. and Boguraev, B. K. [1989] "Dictionaries, Dictionary Grammars and Dictionary Entry Parsing," *Proceedings of the 27th Annual Meeting of the ACL*, Vancouver, B.C., June, 1989, pp. 91-101.

Neff, M. S. and McCord, M. C. [1990] "Acquiring Lexical Data from Machine-Readable Dictionary Resources for Machine Translation," IBM Natural Language Processing ITL, Paris, 1990.

Shieber, S. M. [1986] *An Introduction to Unification-Based Approaches to Grammar*, CSLI Lecture Notes No. 4, Center for the Study of Language and Information, Stanford, CA.

Walker, A. (Ed.), McCord, M., Sowa, J. F., and Wilson, W. G. [1987] *Knowledge Systems and Prolog: A Logical Approach to Expert Systems and Natural Language Processing*, Addison-Wesley, Reading, Mass.

Wilks, Y., Huang, X-M., and Fass, D. [1985] "Syntax, Preference and Right-Attachment," *Proc. 9th International Joint Conference on Artificial Intelligence*, Los Angeles, pp. 779-784.

Woods, W. A. [1973] "An Experimental Parsing System for Transition Network Grammars," in R. Rustin, *Natural Language Processing*, pp. 111-154, Algorithmics Press, New York.

On the Logical Structure of Comparatives*

Manfred Pinkal
Universität Saarbrücken

Logic and natural-language analysis are interconnected on two levels. First, logically based representations and deduction techniques have proved to be a widely applicable and very useful device for reasoning about grammatical structure (cf. the "Parsing as Deduction" approach in logical programming (Pereira/Warren 1980), or the unification grammar paradigm, which is narrowly related to order-sorted logic). Second, there is a specific connection on a more substantive level. Logic is not only a convenient tool to represent and process information **about** natural language, but it also provides an adequate formal framework to model the information that is conveyed by natural language itself. Since the Seventies, logic has been successfully used in natural language semantics to represent meanings of natural language expressions in a systematic and precise way. Montague grammar with intensional type logic as meaning representation language was the standard of the Seventies (cf. Montague 1974).

During the Eighties, not only the formalisms in NL semantics have changed, but there have been important changes in the over-all methodology. One of these changes has been that the parsimony with respect to ontological assumptions and lexical structures ("Construct every meaning by set theoretic operations on a minimal basic inventory, and derive the meaning of every word by set theoretic operations on the meaning of its stem") has been given up as a methodological principle, and instead has been replaced by a more cautious use of construction techniques. A nice result of this change was the insight that far more of natural language meaning, than was originally assumed, can be treated in the framework of first-order predicate logic, and, consequently be processed using standard deduction techniques. This is one of the reasons why logical semantics of natural languages has made its way into natural language processing, during the Eighties, and has successively lost its flavor of being a fascinating, but practically useless toy for theoretical linguists. This paper is intended to show that the methodology indicated above can be successfully applied not only in the field of plurals and events, but also leads to a rather simple and natural semantic representations for various kinds of adjective constructions.

---

* The ideas presented in this paper were developed at the university of Hamburg in connection with the project "Graduierung und Komparation" which is sponsored by the Deutsche Forschungsgemeinschaft. The paper was worked out in a period of time which the author spent as a visitor at the Institut für wissensbasierte Systeme, IBM Germany, in Stuttgart.

The design of a language of logic that is appropriate to represent meaning structures in a certain field of natural language is one typical task for natural language semantics. It may appear to be the main task for semantics, from the view of a logician looking at natural language, the mapping of natural language surface strings into logical representations being just a matter of routine. Actually, this second task is far from being trivial, however. The design of construction rules for building up representations may even appear to be the proper task of semantics, and the design of representations just as a prerequisite. Comparatives form a field of phenomena that impressively illustrates this fact. The second part of this paper will indicate some of the problems connected with the constructions of logical representations for comparatives, and sketch some possible solutions. It will also turn out that careful investigation of the construction mechanism allows to draw conclusions on the design of the target representations themselves.

1. Degrees

Examples like (1) suggest a semantic representation of comparatives as two-place relations between objects.

(1)   John is taller than Bill

If this were true, one could accomodate the representation language to the treatment of comparative constructions by just adding standard relational expressions of the form *taller'* or $>_{tall}$, for *tall* and, accordingly, for each gradable adjective. A very cursory glance at the phenomena shows, however, that this view of comparatives - to which I will refer as the "direct comparison" approach in the following - is untenable.

(2)   John is taller now, than he was five years ago

(3)   John is faster on his bike, than he is on his motorcycle

(4)   The desk is wider, than it is high

In sentences (2) - (4), the same person John occurs on both sides of the comparison. One could try counter the difficulty in the case of sentence (2) by assuming that it is not the objects themselves, but rather stages of objects that are compared (John now and John five years ago). However, this proposal does not work very well with sentence (3), and it is clearly hopeless for sentence (4), where properties of the same stage of an object are compared.

The difficulties immediately disappear, if we take degrees to be the arguments of the comparative relation, instead of the objects : (2) - (4) then convey the respective informations that the degree to which John is tall now exceeds the degree to which John was tall five years ago, that the degree to which John is fast on the bike exceeds the degree to which he is fast on the motorcycle, and that the degree to which the desk is high exceeds its width.

What are degrees, and how do they fit into the framework of natural language semantics? Essentially two proposals have been made to specify degrees in terms of more familiar entities. One class of comparative theories (the "positive-based theories", e.g., Kamp 1975 and different accounts based on fuzzy logic) take their outset from the description of the positive, as in (5), using a many-valued or supervaluation-style logic.

(5)   John is tall

Predicative *tall* is a vague predicate that does not come with a precise denotation, but can apply to objects to a higher or a lower degree. Positive-based theories of the comparative take comparatives to be comparisons between degrees of applicability of a predicate. The most straightforward formalization is to express degrees of applicability by numbers taken from the real interval [0,1], which appear as truth degrees on the sentence level, and accordingly, to interpret comparatives as the arithmetic "greater than" relation. Sentence (1), e.g., would come out to be true if the degree of truth assigned to *John is tall* exceeds the degree of truth of the predication *Bill is tall* .

The method yields truth conditions for sentences (2)-(4). Unfortunately, it does not always provide the intuitively correct truth conditions. John may be very fast on his bike (say, to degree 0.9), and Bill may be pretty slow on his motorcycle (to degree 0.3), but sentence (3) may be false, though, at least on one reading, since John has no chance to achieve the same speed, by bike, as Bill with his motorcycle. One may object here that the semantics of degree adjectives is highly context-dependent. In order to assign the predication (5) a degree, one has to know, among other things,with respect to which comparison class the predicate *tall* is used in the context of its utterance (Klein (1980) presents a theory of adjectives where this idea is worked out in detail). The assumption that comparatives require identical contexts for the compared predications may help to repair some of the problem cases. E.g., we might read *fast* on both sides of the comparison uniformly as "fast on a vehicle". However, it is hard to imagine how the comparison class should look like that brings about the appropriate truth conditions for sentence (4). The truth conditions for this "interdimensional comparative" are intuitively clear: (4) is true, if and only if the width of the desk exceeds its height. The question to which degree *The desk is high* and *The desk is wide* are true as independent sentences does not play any role:

Usually, also a very high desk is not higher than it is wide. For this and some other reasons (discussed in Pinkal (forthcoming)), degrees of truth or applicability of a property cannot serve as the ontological foundation for degrees in general. A height degree cannot be taken to be a measure of the extent to which the property expressed by the predicate *high* applies, a height degree rather is a property of objects. Comparatives are relations on this more concrete and specific kind of degrees.

The argument against truth or applicability degrees is mainly based on evidence from interdimensional constructions, and thus it might be restricted in its validity to "dimension adjectives" like *high, wide, fast, heavy* . Comparatives of "evaluation adjectives" like *good, beautiful, intelligent* (the distinction as well as the argument in favor of a differentiated view are due to Bierwisch 1987) - may refer to degrees on the more abstract level. For "metacomparatives" like (6) this level   must be available, anyway.

(6)   John is more beautiful than he is intelligent

Metacomparatives are a pretty marginal type of construction, however. For the primary readings of comparative constructions with dimension adjectives, we must assume degrees which have their places on scales of height, speed, weight, etc.

The second proposal for an integration of degrees into a formalism for natural language semantics stems from Cresswell (1976), and is taken up again in Klein (forthcoming). The idea is to take the degrees involved in comparative constructions to be specific degrees of height, weight, etc., but to reconstruct the degree concept on the basis of other concepts already available in the ontology. In short, the reconstruction runs as follows: Model structures with universe U are assumed to contain a two-place pre-ordering relation $\geq_A \subseteq U \times U$ for every adjective A ($\geq_A$ reflexive, transitive, connected). $\geq_A$ can be split up, in the usual way, into a "greater" part $>_A$ and an "equal" part $\sim_A$:

(7) $a \sim_A b := a \geq_A b$ and $b \geq_A a$
       $a >_A b := a \geq_A b$ and not $a \sim_A b$

Since $\sim_A$ is an equivalence relation, it can be used to construct the domain of degrees for A as a partitioning of the universe into sets of individuals which are equal with respect to A , which is done in (8).

(8)   The domain of degrees for A :

       $D_A := \{ [a]_{\sim_A} \mid a \in U \}$

Next, an ordering relation $\succeq_A \subseteq D_A \times D_A$ (having all properties of $\succeq_A$ plus anti-symmetry) is defined on the basis of $\succeq_A$, in (9).

(9)   For d, d' $\in D_A$:

   $d \succeq_A d'$ iff for some $a \in d, b \in d'$: $a \succeq_A b$

The definition is consistent, since the $\succeq_A$ relation holds or does not hold independent of the choice of a and b.

Technically, the reconstruction is completely standard. Intuitively, it yields adequate results for the cases which also work with direct comparison, like (1). It leads, however, to similar difficulties with examples (2) and (3), as the positive-based theories: There is no unique "taller" or "faster" relation between individuals as such, but at the best between individuals at certain times or individuals in situations of certain types. For interdimensional comparatives, the reconstruction approach fails completely: There is a relation $\succ_{high}$ defined for D_{high} , as well as a relation $\succ_{wide}$ on D_{wide} , but no $\succ$ relation is defined across domains, as we would need it to interpret sentences like (4). We could try to repair the situation by by adding an equivalence relation to our universe which associates those degrees of height, width, length, etc., with each other which denote the same distance. But then why shouldn't we just take the distances themselves to be degrees, instead of working with complex constructions, which have to make reference to the concept of distance anyway, in some way or the other?

To sum up, the positive-based as well as the reconstruction approach to relate degrees to other concepts leads to inadequate predictions or to considerably complex and unnatural constructions, in at least some important cases. To be sure, degrees cannot be simply taken to be numbers, either. Numbers are assigned to degrees by measurement functions in a basically arbitrary way (Of course, measurement functions have to obtain certain restrictions in order to be consistent; for an account of measurement, see Krantz et al. 1971).

Therefore, this paper pursues an alternative strategy: Degrees are represented as a sort of basic entities in their own right. There are several facts supporting this decision. Degrees can be referred to by names (e.g., *1.80 m* or *75 kg*), by descriptions (*the height of the desk, the weight of Bill*), and by means of deixis (*that tall* , accompanied by an appropriate gesture), as ordinary objects can. And there is a considerable gain in simplicity and plausibility for the modelling of comparative information. E.g., since degrees are independent from adjectives, different adjectives can be related to the same domain without problems, which explains the phenomenon of interdimensional comparatives; on the other hand, since degrees are not generally applying degrees of truth or applicability, it is plausible why interdimensional constructions do not work for arbitrary pairs of adjectives.

There are several reasons in favor of independent domains of degrees, and little that speaks against the assumption, except a principal disagreement with the augmentation of the ontology. Let us therefore assume a structure of the universe as indicated in (10), which consists of standard objects and degrees as basic sorts of entities, where the domain of degrees is further subdivided into classes internally connected by ordering relations.

(10)

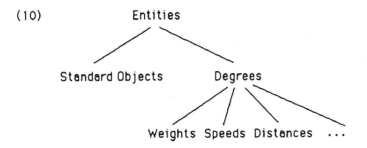

I will refer to these ordered classes as "scales". Note that scales are not meant to be metric scales. It is not even required that they are capable of metrization, although in most cases they will be, in one or the other way.

Finally, let me mention that the treatment of comparatives by an independent degree concept is not an original idea. Bierwisch 1987, e.g., has worked out this idea, too. His approach aims at different phenomena. Thus, he comes to another and much more complex formalism than the one proposed in this paper.

2. Adjectives, Dimensions, Scales

Linguistic work on comparatives often starts out from a naive view of degrees, assuming that each adjective brings its own domain of degrees with it. Thus, we encounter degrees of largeness, heaviness, height and width, as well as degrees of smallness, lightness, and shortness. As the discussion of interdimensional comparatives has shown, such a one-to-one correspondence cannot be held upright: *high, wide,* and *tall* refer to the same domain, i.e. the scale of distances. *heavy* and *light* also refer to the same scale, as the members of many antonym pairs do. There is a difference, however, between the way how the members of the field of spatial dimension adjectives are related to each other, and the semantic relation within antonym pairs : *high, wide,* and *tall* make reference to the same scale, but are otherwise semantically independent, since they are concerned with different dimensions. *heavy* and *light* not only refer to the same scale, but also to the same dimension. The possible types of the interrelation between scales, dimensions, and adjectives are graphically indicated in (11). To complete the picture, another type of relation is added. An adjective like *large* may refer to different dimensions and scales, e.g., either to the area or to the population, if it is applied to cities.

(11)

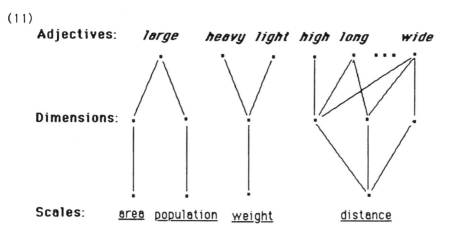

Some comments are in order about the different types of interrelations presented in (11). The antonymy relation seems to be the most straightforward and systematic case. *light* and *short* refer to essentially the same scales as *heavy* and *tall* , respectively, they just use the inverse of the respective $\geq$ -relations. This explains the equivalence of (12) and (13).

(12)   John is taller than Bill
(13)   Bill is shorter than John

However, as the evidence presented in Bierwisch (1987) shows, there is more about the antonymy relation than just inversion.

(14)   John is as tall as Bill

(15)   Bill is as short as John

The equative constructions (14) and (15) should be equivalent. But (15) carries the presupposition that Bill and John are short, whereas (14) does not. Also, *1.80 m tall* is fine, *1.80 m short* , on the contrary, is clearly an unusual and marked construction. Bierwisch draws rather wide-reaching conclusions from these observations, which I will not take over for different reasons. One should realize, however, that there are non-obvious and non-trivial problems connected with the mapping of adjectives onto dimensions.

The situation is yet more complex in the field of spatial dimension adjectives. A sophisticated analysis of this area is presented in Lang (1987). I can only give a rough outline of the central thoughts, here. The basic idea is that "object schemata" are attached to different kinds of objects, which may distinguish in the number of "desintegrated" object axes (bricks and buildings have three of them, balls have none), and in certain features which are inherent to the object axes. E.g., a pole has a distinguished "maximal" ax, a building in general a "vertical" ax, a tower has a maximal and a vertical ax which coincide. The selection of a dimension (object ax) for an adjective in a given predication is done by a complex algorithm, which takes as parameters the object schema of the argument as well as information about the objects current position. E.g., *long* said of a pole always refers to the maximal ax, *high* can be used to refer to the same ax, but only, if the pole stands in upright position. The maximal ax of a brick can be alternatively referred to with the adjectives *high* , *long* , and *wide* , according to the brick's position. To sum up, height, length, width are not spatial dimensions of objects by themselves; they are linguistic concepts which are related to dimensions by a complex and context-dependent mapping process.

Finally, let us look at the *large* example. A pretty good approximation to the semantic function of *large* is to say that it picks out some prominent extensive property of the argument. This property can be ground or surface area for spatially extended objects (like rooms and lakes), number of individuals for social objects (like faculties and birthday parties), memory capacity for computers, etc. *large* applied to cities is ambiguous because cities are spatially extended objects and social institutions, at the same time. Similarly, *good* (in one basic reading) refers to the proper function of its argument object, and *clever* to some skill of its argument person (the notorious ambiguity of *clever* being due to the fact that people usually have several skills, at the same time). The way adjectives like *large* , *good* , and *clever* select dimensions is important, but probably rather a matter of extensive conceptual modelling than of one precise algorithm.

In this section, I have focussed on the upper part of Figure (11), giving an informal account of the relation between adjectives and dimensions. In the next section, I will concentrate on a formal elaboration of the lower part of (11), specifying a model structure concept which incorporates the concepts of dimension and scale.

3. The Representation of Degree Information

According to the informal considerations of sections 2 and 3, the requirements for a degree ontology can be summarized as follows: First, we need <u>scales</u> as ordered domains of degrees. Second, we need a formal representation of the concept of <u>dimension</u>.

Functions from the domain of standard objects to scales are a good candidate. Third, we have to model interdimensional comparatives like (4). This can be achieved by representing different spatial dimensions as different functions with identical range.

The model structure definition given below, will implement the concepts of scale and dimension in a slightly different manner, using a dimension specific two-place comparison relation across domains as its basis. This relational formalization is basically equivalent to the "dimensions as functions" approach sketched above (set aside some details which have been left out in the last paragraph), and could in principle be replaced by the latter. One reason why I prefer the relational approach is that a comparison relation appears to be a very general scale-independent ingredient of many adjectival and non-adjectival concepts, which I have pointed out in my work on vague predicates (cf. Pinkal (forthcoming)). E.g., the predicate *chair* has a two-place relation ("at least as much of a chair as") as part of its semantic representation, which is needed to model context change relations between vague concepts. Gradable adjectives (maybe, only the dimension adjectives; cf. Section 2) form that special case of vague predicates where, in addition, scales are involved. The representation of dimensions as functions to some extent obscures this relationship. Another motivation for the relational representation is that it supports a straightforward and simple interpretation of different kinds of adjectival constructions; this will become visible below.

Before defining the model structure concept, let me give a first outline of the representation language which I intend to employ. I assume a two-sorted language of first-order logic with

-- constants and variables for standard objects

(john, bill, the-desk, . . . ; x, y, z, . . .)

-- constants and variables for degrees

(1.80m, 77kg, 130km/h, . . . ; d, d', d_1, d_2, . . . ; I will not deal with the specific problems connected with measurement, here; for a discussion of measurement problems with adjective constructions see Klein (forth-coming))

-- n-place standard predicates for some natural numbers n

(man, woman, love, give, . . .)

-- degree predicates (about which I will become more specific later).

(16)   A Model Structure with Degrees for a first-order language L with degree
predicates is a 6-tupel

M = < U, D, V, Δ, Dom, $\geqslant$ > , where

(i)   U, D $\neq \varnothing$, U $\cap$ D = $\varnothing$

(the disjoint domains of standard objects and of degrees,
respectively)

(ii)   V is a function which assigns every standard-object constant of L a
member of U, every degree constant of L a member of D, and every n-place
standard predicate of L a subset of U^n

(iii) $\Delta \neq \varnothing$ (the set of dimensions)

(iv)   Dom is a family of sets $Dom_\delta \subseteq U \cup D$ for each $\delta \in \Delta$

(the domains for the respective dimensions;in the following, "U_δ" and
"D_δ" are used for $Dom_\delta \cap U$ and $Dom_\delta \cap D$, respectively)

(v) $D_\delta = D_{\delta'}$ or $D_\delta \cap D_{\delta'} = \varnothing$, for $\delta, \delta' \in \Delta$

(vi) $\geqslant$ is a family of reflexive and transitive 2-place relations
$\geqslant_\delta \subseteq Dom_\delta \times Dom_\delta$

(in the following, $>_\delta$ and $\sim_\delta$ are used, according to definition (7), as the
"greater" and "equal" part of $\geqslant_\delta$, respectively)

(vii)   for d, d' $\in D_\delta$: if d $\sim_\delta$ d', then d = d'

(viii)   for a $\in U_\delta$, there is exactly one d $\in D_\delta$ such that a $\sim_\delta$ d

(ix)   for $\delta, \delta' \in \Delta$: if $D_\delta = D_{\delta'}$, then for all d, d' $\in D_\delta$:
d $\geqslant_\delta$ d' iff d $\geqslant_{\delta'}$ d'

Conditions (i) to (iii) need no further comment. Conditions (iv) and (v) specify the
domains of the dimension specific comparison relations $\geqslant_\delta$: Each dimension δ comes
with its particular domain of degrees D_δ. As stated by (v), the domains can be identical
for different dimensions (they actually are identical in the spatial case); between different

dimensions, there is no overlap. Also, dimensions go with certain kinds of objects only, which are specified by the U_δ sets of (iv): A "head-count" dimension makes only sense for collective objects, a "third object ax" dimension only for objects which are three-dimensional (and have three desintegrated object axes, in the sense of Lang's account).

Conditions (vi) - (viii) specify the properties of the $\geqslant_\delta$-relations: They relate standard objects to standard objects, degrees to degrees, and standard objects to degrees. $\geqslant_\delta$ is required to be reflexive and transitive, only; the ordering can be partial, since their may be incomparable objects with respect to certain dimensions (consider the case where *big* is used as a comparison of gestalt properties of different individuals: Is a rhino bigger than a giraffe?). A careful consideration of dimension properties may lead to the result that $\geqslant_\delta$ not even is transitive. The standard evidence in favor of this assumption is the "Sorites" or "heap paradox". I do not want to pursue the argument, here, and refer to Pinkal (1985) for a discussion. Condition (vii) makes $\geqslant_\delta$ a (partial) ordering relation on its scale; for standard objects, it is a pre-ordering relation only. Because of (viii), $\geqslant_\delta$ specifies a function from U_δ to D_δ. Finally, (ix) requires two dimension relations to coincide on the same domain of degrees, which is necessary for a consistent interpretation of interdimensional comparisons.

Having specified the model structure concept, let us now supply that part of the representation language L responsible for degree information, which I had left out above. I will assume that the language contains two sets of two-place degree predicates - the positive and the comparative degree predicates - such that there is a one-to-one correspondence between the members of the two sets (*tall, high, heavy, . . . ; taller, higher, heavier, ...*). Both kinds of degree predicates take a standard-object term as their first argument, and an arbitrary term as their second argument. This completes the inventory of constants of L.

The next task is to specify assignment and interpretation concept for L on the basis of the model structures defined in (16). A variable assignment h assigns standard objects of M to standard-object variables and degrees of M to degree variables. Degree predicates are the only non-logical constants which are not provided with a fixed denotation by M. The idea is to assign a positive predicate like *tall* the $\sim$ part and the comparative predicate *taller* the $>$ part of the corresponding $\geqslant$ relation. As shown in Section 3, however, there is no such thing as the corresponding $\geqslant$ relation: The dimension which a degree predicate refers to is dependent on different aspects of the context of utterance. To model this property, we could, e.g., assume a special two-place function ϕ first which maps pairs of contexts and adjectives onto dimensions, and then define the context-specific assignment V_c of relations to degree predicates, as in (17).

(17) $V_c(tall) = \sim \phi(c, tall)$

$V_c(tall) = \; > \phi(c, tall)$

Except for the context parameter which is needed for degree predicates, the interpretation function $[[A]]_{M,h,c}$ of complex expressions A of L is first-order standard. The truth conditions for one atomic predication are calculated in (18), for illustration.

(18) $[[taller(john, 1.80m)]]_{M,h,c} = 1$

iff $<[[john]]_{M,h,c}, [[1.80m]]_{M,h,c}> \in [[taller]]_{M,h,c}$

iff $V(john) \sim \phi(c, tall) \; V(1.80m)$

In the rest of the paper, I will disregard the complex contextual relations I have discussed in Section 3, since they do not interfere with the matters to be discussed, and use the incorrect, but simpler notations $\sim_{tall}$ and $>_{tall}$, respectively, to refer to the denotations of the degree predicates *tall* and *taller*.

In the following, some examples for adjective constructions are listed (under (a)), together with their logical representations (b) and the truth conditions assigned to them by the interpretation function (c). **j** and **b** stand for the persons John and Bill, **t** for the desk, and **d** for the degree denoted by "1.80m". Also, I employ the Russellian description operator ι, for abbreviatory purposes.

(19)   (a)   John is taller than Bill
       (b)   taller(john, bill)
       (c) $\mathbf{j} >_{tall} \mathbf{b}$

(20)   (a)   John is taller than 1.80m
       (b)   taller(john, 1.80m)
       (c) $\mathbf{j} >_{tall} \mathbf{d}$

(21)   (a)   John is taller than the desk is wide
       (b)   taller(john, ιd: wide(the-desk, d))
       (c)   there is $\mathbf{d} \in D: \mathbf{j} >_{tall} \mathbf{d}$
             and **d** is the only **d** such that $\mathbf{t} \sim_{tall} \mathbf{d}$

(22) (a) John is as tall as Bill
    (b) tall(john, bill)
    (c) $\mathbf{j} \sim_{tall} \mathbf{b}$

(23) (a) John is 1.80m tall
    (b) tall(john, 1.80m)
    (c) $\mathbf{j} \sim_{tall} \mathbf{d}$

(24) (a) John is as tall as the desk is wide
    (b) tall(john, ιd: wide(the-desk, d))
    (c) there is $\mathbf{d} \in$ D: $\mathbf{j} \sim_{tall} \mathbf{d}$
       and $\mathbf{d}$ is the only $\mathbf{d}$ such that $\mathbf{t} \sim_{tall} \mathbf{d}$

As (19) to (24) show, the proposed semantic account of adjectives leads to simple and high uniform representations and interpretations. The result is due to the fact that at several points decisions have been made which are different from the usual way comparative theories are designed, in a certain amount of redundancy in representation preference over the use of reductions.

First, direct comparisons (as (19) and (22)) are treated as constructions in their own right, in parallel to measure constructions ((20),(23)) and as less complex cases than sentential comparatives ((21),(24)), on all levels of analysis. In most theories, NP comparatives are considered to be elliptic sentential comparatives, as indicated in (25a), and accordingly represented in a way similar to (25b) (see e.g. Cresswell 1976, v. Stechow 1984, Bierwisch 1987; exceptions are Hoeksema 1983, Heim 1985).

(25) (a) John is taller than Bill (is tall)
    (b) taller(john, ιd: tall(bill, d))

The problems that these "ellipsis theories" have to face are first a reconstruction step on the way from the syntactic surface to the logical representation for which no satisfactory general description exists, and second, a target representation that somehow appears too complex (Evidence beyond the mere impression of inadequacy is possibly provided by some kinds of modally embedded comparatives; see Pinkal forthcoming).

Second, I have decided to provide two lexical representations for each adjective, a positive and a comparative one (after all, a superlative entry must be added, too, and perhaps also a separate entry for the one-place positive, as in *John is tall*). There has been a strong tendency towards avoiding these multiple representations in the lexicon, by reducing one of two morphological forms to the other. v. Stechow (1984) makes the most plausible reduction proposal I know about, roughly along the following lines:

(26)　(a)　John is 10cm taller than Bill (is tall)

　　　(b)　tall (john, ιd: tall(bill,d) + 10cm)

(27)　(a)　John is taller than Bill (is tall)

　　　(b)　∃d' tall (john, ιd: tall(bill,d) +d')

The basic construction type is the positive measure phrase ("tall(x,d)") The function of the comparative morpheme -er is to transform the two-place degree predicate into a three-place predicate which is used in"differential comparatives" (26). If the third argument (the difference) is not explicitly mentioned, as is the case in standard comparatives like (27), the argument position is bound by existential quantification. The advantage of this account is that it does not only reduce the comparative to the positive degree predicate, but also integrates differential constructions. One of the disadvantages is that the interpretation of the conceptually simple direct comparative becomes a pretty complicated matter, which requires ellipsis reconstruction, implicit arguments, and assumptions about the structure of degree domains (which must allow an addition operation in order to make comparatives possible at all). The alternative treatment described in this paper takes different morphological forms (like positive and comparative) as lexical entities of equal status, which are semantically related in a systematic way: Both refer to the same dimension(s); they differ in that the positive picks out the ~ part and the comparative picks out the > part of the dimension-specific comparison relation. This way of description has an analogy in the theory of plurality proposed by Link (1983): All common nouns refer to the same domain (the semi-lattice of sums). However, the have partitioned the domain in that singulars refer to atoms, whereas plurals refer to non-atomic members.

A third aspect in which the proposed account differs from most existing comparative theories, will start the discussion of the next section.

4. Syntactic Structure and Logical Representation

The analysis proposed in this paper as widely as possible saves the parallelism between surface constituent structure of comparative constructions and semantic representation. I will try to make this point clear by contrasting my analysis (and v.Stechows, which is in this respect similar) with a type of analyses which I will refer to as the discontinuous ones, for reasons which will become clear in a moment. v.Stechow assigns stem and suffix of a comparative form different semantic functions. However, the semantic function of the suffix can be described as a local one, taking a word meaning and making it another word meaning. Theories like that of Cresswell (1976) and Heim (1985) separate stem and suffix in a more radical way. The -er -suffix is supposed to have a function similar to a noun phrase determiner. It takes a sentential complement and forms a term phrase (a generalized quantifier), which in turn takes the degree predicate denoted by the remainder of the matrix sentence as argument. I will sketch Heim's analysis in the following. The syntactic structure for (28) which is the input for the construction of a meaning representation is indicated in (29); (30a) is the representation resulting from the translation of the constituents of (29), and (30b) the predicate logic formula resulting from λ-conversion. The notation is standard type-logic notation from Montague grammar. The use of second-order variables (like D as a variable over one-place degree predicates) is ontologically harmless, since they disappear by λ-conversion. The translation of the er-than -Term of (29) can be paraphrased as "a degree which exceeds the degree to which the desk is wide".

(28)   John is taller than the desk is wide

(29)   [$_S$ [-er-than [$_S$ the desk is wide]]$_d$ [$_S$ John is d-tall]]

(30)   (a) $\lambda d_2 \lambda D \, \exists d_1 \, (D \, (d_1) \wedge d_1 > d_2)$
             $(\iota d\!:\!wide \, (the\text{-}desk, \, d))(\lambda d\!: \, tall(john, \, d))$

        (c) $\exists \, d_1 \, (tall \, (john, \, d_1) \wedge d_1 > \iota d \, wide(the\text{-}desk, \, d))$

The extraction of the er-than -Term in (29) from its sentence-internal position is a standard technique known as "quantifier raising".It is employed to give the corresponding quantifier wide scope over the whole construction; there are other techniques to bring this scopal effect about, but something of that kind is necessary for many applications, and I will use quantifier raising,too, in the following, since it makes no differences for the purposes of this paper.

The problematic aspect of the analysis consists in the very assumption of an er-than - Term, for it implies that the comparative suffix is an independent, scope-bearing syntactic entity, which may have its location in arbitrary distance from its stem. There is no single part in the logical representation of (28) which corresponds to the adjective phrase taller, than the desk is wide or even to the word taller , and this is the worse because taller intuitively has a simple and unproblematic conceptual correlate. The discontinuous analysis has no possibility to express this concept, but subdivides it from the beginning into one part providing the dimension, and a second part consisting of the >-relation.

For the sake of completeness, let me give an example how Heim models direct comparison. Here, the input to meaning construction for sentence (31) is (32), where -er again functions as an independent operator, taking in this case a pair of objects and a function from objects to degrees as arguments, and leading to meaning representation (33).

(31)   John is taller than Bill

(32)   -er(<john, bill>, λx ιd: x is d-tall)

(33)   ιd: tall(john,d) > ιd: tall(bill,d)

Let us now look at the analysis argued for in this paper: In all comparative constructions considered, *taller* is a basic two-place predicate, with the only special feature that the second argument comes with the particle *than* and may be a degree-denoting phrase: either a measure constant or a degree term expressed by a sentential complement. The syntactic structures which are presupposed by the analysis, are indicated in (32), (33), and (34).

(32)   [s [NP John] [VP is [AP taller than [NP Bill]]]]

(33)   [s [NP John] [VP is [AP taller than [NP 1.80 m]]]]

(34)   [s [NP John] [VP is [AP taller than [s the desk is wide]]]]

(32) and (33) straightforwardly lead to the corresponding meaning representations (19b)and (20b). For (21b), we have to admit quantifier raising, corresponding to Heim's analysis in (29). What is raised is however not the artificial *er-than-* Phrase construct, but rather the term-denoting complement sentence.

There is one gap left in the derivation of semantic representations from the syntactic surface. Up to now, I did not care about the question how the term denoted by the complement of sentential comparatives comes about. It is commonly accepted that the degree position in sentential complements is somehow bound from outside, the complement thus having an interior structure like [s [NP the desk] [vp is [AP d wide]]]. Syntactic evidence for that analysis is provided by the ungrammaticality of sentential complements with an overt degree phrase (*than the desk is 1.80m wide, *than the desk is very wide). Aside from this fact, the syntactic analyses diverge, and the semantic type of the complement (a definite description, as in Heim's and v.Stechows analysis, or a λ-expression denoting a degree predicate) is stipulated. Also, the definite description analysis turns out to be inadequate, at closer inspection.

(35)   John is taller than Bill or Fred are

The straightforward analysis of the complement of sentence (35) as ιd:Bill or Fred is d tall would render a true sentence only, if Bill and Fred are of identical height (otherwise, the uniqueness condition is not satisfied, and representation (36) becomes false). Quantifier raising of *Bill or Fred* will give the connective "∨" widest scope, and by that result in the reading noted in (37) which says that John is taller than the shorter of the two persons Bill and Fred.

(36)   taller (john, ιd: (tall(bill,d) ∨ tall(fred,d)))

(37)   taller (john, ιd: tall(bill,d)) ∨ taller (john, ιd: tall(fred,d))

(35), however, has a reading, which by the way seems to be the preferred one, saying that John is taller than the taller of the two persons Bill and Fred. The example might appear rather far-fetched. The same semantic effect can however be observed in all those cases where the complement admits of alternative degree assignments, as examples (38) - (40) demonstrate.

(38)   John is taller than <u>any</u> student is

(39)   John is taller than Bill <u>ever</u> was

(40)   John is richer than a professor <u>can</u> be (rich).

Examples (38) and (39) assume *any* and *ever* to be negative polarity items with existential force, which seems to be a cogent assumption at least in the case of *ever*. Interpretations of (38) - (40) along the lines of (36) would have the absurd implication that all students are of equal height, Bill did never grow, and a professor can have one and only one income, or the much too weak meaning that John is taller as the shortest student, taller than the least height Bill ever had, and richer than the poorest possible professor. Instead, (38) to (40) definitely mean that John exceeds the tallest student in height and the richest possible in income. Arnim v.Stechow proposed to solve the problem by the stipulation that the ι-operator, when used in comparatives, denotes not the <u>only</u> degree, but rather the maximum degree satisfying the description. This provides appropriate results for examples (35)-(40). (For an extensive discussion of the problem cases see v.Stechow 1984 and Pinkal forthcoming). I want to point out in the following that there is a yet more plausible analysis of the semantics of comparative complements which immediately follows from a certain syntactic analysis.

Chomsky (1977) proposes to analyze sentential comparative complements as a kind of free relative sentences, the degree variable being bound by a relative pronoun, as indicated in (41).

(41) $[_S (wh_d) [_S$ the desk is d wide]]

In English, the pronoun is phonetically empty. In other languages, however, it can be realized (as in the Scandinavian languages, see Hellan 1981) or even must be realized(as in Italian).

Standard free relative sentences have a strong tendency towards a universal reading, cf. example (42).

(42)   Who has finished the test, may leave the class-room

Thus, it seems plausible to analyze sentential complements as terms, for syntactic reasons, however not as definite descriptions, but as universal terms.

The syntactic structure of (28), according this analysis, is indicated in (43), with the complement in raised position. Given *taller* as a two-place predicate, and the complement as a free relative with universal semantics, we arrive at translation (44a) which by λ-conversion reduces to (44b), without any further specific assumptions.

(43) $[[_{S'} (wh_{d'}) [_S [_{NP}$ the desk] $[_{VP}$ is $[_{AP}$ d' wide]]] $]_d$

   $[_S[_{NP}$ John][$_{VP}$is[$_{AP}$taller than d]

(44)   (a) $\lambda D\ '\lambda D\ \forall d_1 (D\ '(d_1) \rightarrow D\ (d_1)) (\lambda d'\ wide(the\text{-}desk, d'))$
                                    $(\lambda d\ taller\ (john, d))$

   (b) $\forall d_1 (wide(the\text{-}desk, d_1) \rightarrow taller\ (john, d_1))$

(45) $\lambda D\ \forall d (tall(bill,d) \lor tall(fred, d)\ \rightarrow D\ (d))$

The complement of sentence (35) gets accordingly the representation (45), which can be paraphrased as "all degrees d such that Bill or Fred is d-tall", and gives the reading that John has to be taller than every of the two persons. Accordingly, the adequate readings for (39)-(41) result. There are even cases where the universal analysis weighs out the modified description analysis of v. Stechow:

(46)   2 is greater than every lower rational approximation of the square root of 2

(47)   John is as tall as Bill or Fred is

In v.Stechows analysis, (46) would be false, since there is no maximal value for the complement. The proposed universal analysis yields true, in accordance with intuitions about the example. The equative would get as one reading that John is taller than the taller of the two persons Bill and Fred, a reading which is intuitively not available, as far as I can see. My analysis predicts that (47) either means that John is as tall as Bill or as tall as Fred (wide-scope or), or that it implies identical height of Bill and Fred.

5. Concluding Remarks

As shown in this paper, we get a uniform, surface-near and intuitively plausible semantic analysis of different adjective constructions of English, on the basis of a few decisions which are along the line semantic research has taken in the last years (broad ontological basis, lexical representation of morphosemantic correspondences) and a thorough semantic investigation of structures which keeps an eye on syntactic facts. I want to conclude the paper by briefly pointing to some open problems. This also gives me the opportunity to do justice to some of the authors the theories of whom I have mentioned in my paper, since the decision for the simple solution usually ruled out the description of phenomena which motivated the more complex approaches. I remind of the open problems mentioned in Section (3), and will restrict myself here to two important problem cases concerning the construction of semantic representations.

First, the decision for the direct analysis and the representation of comparatives as basic predicates makes it difficult to deal with differential comparatives as (48).

(48)   John is 10cm taller than Bill

v. Stechow can treat them in a very elegant way. As I have argued, the trade-off for his treatment is too high: He must make very special assumptions about adjectives and comparatives in general, which I would hesitate to accept. An analysis on the basis of the account presented in this paper must still be found, however.

Second, the treatment of phrasal comparatives by direct comparison and of sentential comparatives by degree term leaves out an important group of comparatives, i.e., phrasal comparatives in attributive use, like (49).

(49)   John has a faster car than Bill

Here, the complement denotes the person Bill, but the comparison is not direct as a comparison of persons, but a comparison of cars to which the persons denoted by the complement and the correlate phrase stand in the ownership relation. Irene Heim's account of phrasal comparatives, which I have given an example for in (31) - (33), aims at this type of construction. (49), e.g., would get the representation (50), which in turn leads to the truth-conditions that the speed of John's car exceeds the speed of Bill's car.

(50)   -er(<john, bill>, λx ιd: x has a d-fast car)

Heim's account, however, is too expensive in the respect that it presupposes a intermediate syntactic structure which is very far from surface, as well as an independent syntactic status of the comparative suffix. Since it moreover leads in certain cases to inadequate results, I cannot accept it as the theory for phrasal comparatives. For attributive cases like (49) there is just no satisfactory analysis available, yet. Comparatives are still one of the most intricate fields for logically based natural-language analysis.

References

Bierwisch, M.: (1987): Semantik der Graduierung In: Bierwisch/Lang (1987)

Bierwisch, M./Lang, E. (eds.) (1987): Grammatische und konzeptuelle Aspekte von Dimensionsadjektiven. Berlin:Akademie-Verlag

Chomsky, N. (1977):, On wh-movement. In: P. Culicover, T. Wasow, A. Akmajian (eds.): Formal Syntax. New York: Academic Press.

Cresswell, M. (1976): The semantics of degree. In: B. Partee (ed): Montague Grammar. New York: Academic Press

Heim, I. (1985): Notes on Comparatives and Related Matters. Ms. Austin, Texas

Hellan, L. (1981): Towards an Integrated Analysis of Comparatives. Tübingen: Narr

Hoeksema, J. (1983): Negative Polarity and the Comparative. In: NLLT 1, S. 403-434

Kamp, H. (1975): Two Theories about Adjectives. In: E. Keenan (ed): Formal Semantics for Natural Language. Cambridge

Klein, E. (1980): A Semantics for Positive and Comparative Adjectives. Linguistics and Philosophy 4, S. 1-45

Klein, E. (forthcoming): Comparatives. In: v. Stechow/Wunderlich (eds.): Handbook of Semantics.

Krantz, D. et al. (1971): Foundations of Measurement I. New York: Academic Press

Ladusaw, W.A. (1979): Polarity Sensitivity as Inherent Scope Relations. In: PhD diss. Austin, Texas

Lang, E. (1987): Semantik der Dimensionsauszeichnungen räumlicher Objekte. In: Bierwisch/Lang (1987), 287-458

Link, G. (1983): The Logical Analysis of Plurals and Mass Terms. In: Bäuerle, R. et al.: Meaning, Use, and Interpretation of Language. De Gruyter: Berlin

Montague, R. (1974): Formal Philosophy. Selected Papers. New Haven: Yale University Press

Pereira, F./ Warren, D. (1980): Definite Clause Grammars for Natural Language Analysis. Artificial Intelligence 13, 231-278

Pinkal, M. (1985): Logik und Lexikon: Die Semantik des Unbestimmten. De Gruyter, Berlin

Pinkal, M. (1990): Die Semantik von Satzkomparativen. GuK Working Paper 1. Hamburg

Pinkal, M. (forthcoming): Imprecise Concepts and Quantification. In a volume edited by J. van Benthem and R. Bartsch. Dordrecht: Reidel

Pinkal, M. (forthcoming): A Discourse Semantics for Imprecise Concepts. In: J. Wedekind/ C. Rohrer (eds.): Unification Grammar and Discourse Theory. Dordrecht: Reidel

von Stechow, A. (1984): Comparing Semantic Theories of Comparison. In: Journal of Semantics 3: S. 1-77

Aspects of Consistency of Sophisticated Knowledge Representation Languages

Udo Pletat

IBM Germany

Scientific Center

Institute for Knowledge Based Systems

Project LILOG

P. O. Box 80 08 80

D-7000 Stuttgart 80

West - Germany

Abstract

The growing interest and the importance of knowledge based systems increase the demand for knowledge representation formalisms supporting the adequate representation of the knowledge to be processed within such systems. The logic-based knowledge representation language L_{LILOG} has emerged from applications in the area of natural language understanding and offers a sophisticated type concept integrating sort description features typically present in languages of the KL-ONE family into the framework of an order-sorted predicate logic. For languages with such rich type systems several consistency problems arise that are not present in neither of the parent languages: order-sorted predicate logic and KL-ONE. We provide a first collection of criteria assuring the existence of models for L_{LILOG} knowledge bases and discuss the relationship between L_{LILOG} and other approaches to sorted knowledge representation languages.

1 Introduction

Intelligent information processing systems are nowadays typically called knowledge based systems. This is due to the idea that intelligent behaviour is based on knowledge about the world. Taking this as the starting hypothesis, it is only natural that the representation of knowledge becomes an essential issue for the development of knowledge based systems.

Looking at the literature on knowledge representation one can observe two major paradigms for the representation of knowledge in general and within knowledge based systems in particular:

- the logic approach, see [Moore 82]

- the semantic network approach, c. f. [Brachman 79]

Choosing a knowledge representation language for the development of a particular knowledge based system should be subject to adequacy considerations. The choice of a concrete language should be preceded by the more important decision according to which paradigm the knowledge shall be represented. The border lines between the different paradigms are quite sharp and the representation aspects addressed by the two paradigms are of rather different nature.

- Predicate logic is quite well-suited for representing knowledge that allows for good axiomatization, see e.g. [Genesereth, Nilsson 87]. A weakness of pure predicate logic is that it doesn't offer a type concept. For the development of large scale knowledge bases the availability of a sort concept has become an accepted language feature since it supports - to some extent - the structuring of knowledge. Order-sorted logic (c.f. [Oberschelp 62], [Walther 87]) overcomes some of the drawbacks of first order logic although the type system inherent to order-sorted logic is often considered too weak.

- The ideas of semantic networks lead to the development of KL-ONE and its derivatives which we will consider here, see e.g. [Brachman 79]. Quite in contrast to what goes on in first order predicate logic, the basic idea of a KL-ONE knowledge base is to define a collection of concepts together with simple relationships between them. We will call the members of the KL-ONE family of languages set description languages due to the in the meantime well-accepted formal semantics definitions for KL-ONE interpreting the syntactic notion of a concept by a set. In the KL-ONE paradigm sorts (= concepts) and their semantic counterpart of sets of objects are emphasized while complex relationships between objects that are typically represented in terms of logial axioms are of lesser interest. The means for representing axiomatic knowledge have been improved in the KRYPTON system (see [Brachman et al. 83]) where first order logic has been integrated into KL-ONE.

  Formalisms with similar basic ideas have been developed quite independently in the area of computational linguistics where feature structure languages have been developed, see e.g. [Shieber 86], [Kasper, Rounds 86], [Bouma et al. 88], [Smolka 88]. The original application of feature structures is to represent grammar rules and lexical entries underlying the parsing process of natural language understanding systems. However, as computer scientists took closer looks at what the computer linguists had developed, the idea that languages for defining feature structures are basically type concepts has become more and more accepted, see e.g. [Beierle et al. 88], [Nebel, Smolka 89].

We introduce an approach to integrate the logic and KL-ONE (we will use the term KL-ONE in the more general sense of set description languages) paradigms for knowledge representation into one formalism. The objective is to arrive at a language offering both a sophisticated type concept according to the ideas of KL-ONE (c.f. [Brachman, Schmolze 85]), STUF (c.f. [Bouma et al. 88]), or Feature Logic (c.f. [Smolka 88]), plus the flexibility of representing axiomatic knowledge we know from predicate logic. Our concern here is to discuss formal properties of this integration emphasizing aspects of consistency, i.e., the existence of models for knowledge bases. The discussion is centered around the definition of (a fragment of) the knowledge representation language L_{LILOG} , see [Pletat, Luck 89].

L_{LILOG} integrates set descriptions in the style of KL-ONE - enriched by disjunction and negation - as a means for describing sorts into an order-sorted predicate logic. As we develop the syntax and semantics of L_{LILOG} we report about several problems of inconsistency L_{LILOG} knowledge bases have to face. It is interesting to observe that the kind of inconsistency we find in L_{LILOG} knowledge bases cannot occur in neither of its parent languages: order-sorted predicate logic and KL-ONE.

The definition of the fragment of L_{LILOG} we consider here proceeds along the lines of defining an order-sorted predicate logic. Thus the major components of a L_{LILOG} knowledge base are a sort hierarchy introducing the universe of discourse, function and predicate symbols over which formulas can be constructed, and of course a set of axioms. In contrast to order-sorted logic, where sorts are just names, we have complex sort expressions in L_{LILOG} . These complex sort expressions are formed due the ideas of KL-ONE and are built from atomic values, features and roles. The sort expressions allow, e.g., to express that sets are the union, or intersection, or complement of other sets. As sort expressions are built from some basic alphabet much in the same way as terms are built over a signature, the sort names, atoms, features, and roles over which sort expressions can be constructed are collected within a sort signature. Together with a collection of sort constraints, where we can enforce that a sort name has to be interpreted by the same set as some sort expression, we obtain sort hierarchies which correspond to the T-Boxes in the KL-ONE terminology, and to the partially ordered set of sorts known in the world of order-sorted logic. The influence of order-sorted logic on atoms, features, and roles is reflected in the fact that we attach sort information to them which is in contrast to what is known from the KL-ONE world:

- atoms are associated to an arbitrary sort; in KL-ONE this may only be the top sort of the lattice of sort expressions

- features may have arbitrary sorts as their domain while the range may be given in terms of an arbitrary sort expression; in KL-ONE both the doamin and the range of a feature have to be the top sort

- roles are relations between an arbitrary sort (first argument) and a set described by an arbitrary sort expression (second argument); in KL-ONE these must both be the top sort

The phenomenon of having inconsistent sort signatures in L_{LILOG} does not occur neither in order-sorted logic nor in KL-ONE. In Section 2 we provide criteria for assuring the consistency of the sort signature of a L_{LILOG} knowledge base.

The story continues with the discussion of sort hierachies of L_{LILOG} knowledge bases. In contrast to both order-sorted logic (where we can always find a universe interpreting the partially ordered set of sorts) and KL-ONE (where T-Boxes are always consistent as long as they don't contain recursive definitions of concepts; also for simple forms of recursive concept definitions the consistency can be assured, see [Nebel 89]) sort hierarchies in L_{LILOG} may be inconsistent as well, even when the underlying sort signature is consistent. This is being discussed in Section 3.

Putting sort hierarchies together with function and predicate symbols in order to form signatures we face again problems of inconsistency unknown to order-sorted predicate logic and KRYPTON, the only member of the KL-ONE family offering a full predicate logic in the A-Box, see Section 4.

The next two sections discuss the formation and semantics of terms, formulas and knowledge bases. As far as consistency is concerned we are getting back to the standard situation of order-sorted logic. Finally, in Section 7 we address some open questions which shall be subject to future research.

2 Sort Signatures and Sort Expressions

A sort signature contains those symbols over which sort expressions serving for the description of sets can be formed. Classical order-sorted logic ([Oberschelp 62], [Walther 87], [Goguen, Meseguer 87]) considers only very restricted sort signatures consisting just of a collection of sort names which are used for describing sets. The language L_{LILOG} has adopted concepts from elaborate set description languages that originated from the KL-ONE family of languages ([Brachman, Schmolze 85]) and languages for defining feature structures which have their origin in the field of computational linguistics ([Kasper, Rounds 86], [Shieber 86], [Bouma et al. 88], [Smolka 88]). According to all these influences on L_{LILOG} , a sort signature contains, in addition to sort names, atoms, features, and roles. In contrast to approaches of the KL-ONE-style and also the feature term languages, we adopt the ideas of sorted logic and tag the atoms, features, and roles of a sort signature with sort information. For the atoms this means that they can be attached to arbitrary sorts, not only to the top sort $\top$ of the lattice of sort expressions as in the KL-ONE style or feature structure languages like STUF or Feature Logic. In this framework, features and roles may have arbitrary source sorts and their target may be an arbitrary sort expression. Again, this generalizes the situation given in feature term languages where the source and target of a feature and a role is always the top sort $\top$.

2.1 Syntax Definitions

This section introduces the syntax of sort signatures and the sort expressions that can be formed with respect to a sort signature. The syntactic constructs introduced in an abstract way will be illustrated by means of examples in an ad hoc concrete syntax which should be self-explaining.

Definition 2.1 *A sort signature* Σ^{SORT} *is a quadruple* $\Sigma^{SORT} = \langle S, A, F, R \rangle$ *where*

- S *is a set of sort names such that* $\{\bot, \top\} \subseteq S$
- $A = \langle A_s \rangle_{s \in S}$ *is a family of atoms*
- $F = \langle F_{s \to se} \rangle_{s \in S, se \in SE(\Sigma^{SORT})}$ *is a family of sets of features*
- $R = \langle R_{s, se} \rangle_{s \in S, se \in SE(\Sigma^{SORT})}$ *is a family of sets of roles*

Sort expressions to be defined in the next step represent complex sorts constructed over the alphabet given by a sort signature.

Definition 2.2 *Let* $\Sigma^{SORT} = \langle S, A, F, R \rangle$ *be a sort signature.*

- *The set of sort expressions* $SE(\Sigma^{SORT})$ *over* Σ^{SORT} *contains*
  - S
  - $se \sqcap se'$
  - $se \sqcup se'$
  - $\neg se$
  - $[a_1, ..., a_k]$
  - **with** f **in** se
  - **with** r **in** se
  - **some** r
  - **agree** fp fp'
  - **disagree** fp fp'

where $se, se' \in SE(\Sigma^{SORT})$, $se \in SE(\Sigma^{SORT})$, $r \in R$, $a_i \in A$ and $fp, fp' \in FP(\Sigma^{SORT})$.

- The family of feature paths $FP(\Sigma^{SORT})$ over Σ^{SORT} is defined as
$$FP(\Sigma^{SORT}) = (FP(\Sigma^{SORT})_{s \to se})_{s \in S, se \in SB(\Sigma^{SORT})}$$
where

  - $F_{s \to se} \subseteq FP(\Sigma^{SORT})_{s \to se}$
  - $f \in F_{s \to s'}$ and $p \in FP(\Sigma^{SORT})_{s' \to se}$ implies $f; p \in FP(\Sigma^{SORT})_{s \to se}$

Before we define the formal semantics of these syntactic concepts, let us give an idea of how formulate taxonomic knowledge with these language constructs.

The basic idea is to describe sets by means of sort expressions. Compared to other set description formalisms like the KL-ONE family of languages, STUF, or Feature Logic, features and roles can have arbitrary source sorts and any sort expression as their target, not only the top sort ⊤ of the lattice of sort expressions. This is a straightforward generalization when embedding these concepts into a sorted logic, since features and roles are (distinguished) one-place functions and two-place relations, respectively.

So we can for example define the sort of vehicles as

>   **sort** *vehicle*
>   **features** *wheels* : [*2, ..., 16*]
>   *doors* : [*0, ..., 4*]
>   *type* : *vehicle-type*

where *bike*, *sedan*, *cabrio*, and *truck* would be atoms of another sort, maybe *vehicle-type*; i.e., we would have another sort declaration

>   **sort** *vehicle-type*
>   **atoms** *bike, cabrio, sedan, truck*

Furthermore, a sort *nat* plus respective atoms modelling the numbers should be contained in the sort signature, i.e. we have

>   **sort** *nat*
>   **atoms** *1, ..., max-nat*

as part of our sort signature.

We now may describe special sets by forming expressions like

>   **with-feature** *wheels* **in** [*4, ..., 16*]

which might describe all kinds of cars, or

>   **with-feature** *type* **in** [*cabrio*]

describing all vehicles having a convertible roof.

Because features are attached to sorts, an expression like

>   **with-feature** *wheels* **in** [*4, ..., 16*]

always describes a set which is a subset of the source sort of the feature occurring in the expression, *wheels* in this case, i.e. the sort *vehicle* subsumes the sort expression **with-feature** wheels **in** [4, ..., 16]. This shows again the generalization of the situation of KL-ONE or Feature Logic: there features are implicitly given as

$$f : \top \longrightarrow \top$$

and the expression

with-feature *wheels* **in** [*4, ..., 16*]

is of course also subsumed by the source sort of the feature, namely by $\top$.

2.2 Semantics Definition

For the syntactic concepts of sort signatures and sort expressions we now provide the semantic notion of universes which are the structures in which we interpret sort signatures and the sort expressions formed over them.

The model-theoretic concept of a universe consists of a collection of sets such that for each sort expression over a sort signature there is a corresponding set interpreting it. Moreover, the sort expressions impose a special structure on the sets being part of the universe of objects. Starting with the interpretation of the sort expressions, we will then say how to interpret the other components of a sort signature. Atoms are elements of the sets interpreting the sort they are attached to; features and roles are interpreted by one-place functions and two-place relations, respectively.

Definition 2.3 *Let* $\Sigma^{SORT} = \langle S, A, F, R \rangle$ *be a sort signature.*
A <u>universe</u> *U for* Σ^{SORT} *is a triple* $U = \langle D, F_U, R_U \rangle$ *where*

- D *is a family of sets* $D = \langle D_{se} \rangle_{se \in SE(\Sigma^{SORT})}$, *the domain of U with*

    - $D_{se} \subseteq D_\top$ *for any sort expression se*
    - $D_\bot = \emptyset$
    - $D_{se \sqcap se'} = D_{se} \cap D_{se'}$
    - $D_{se \sqcup se'} = D_{se} \cup D_{se'}$
    - $D_{\neg se} = D_\top \setminus D_{se}$
    - $D_{[a_1,...,a_k]} = \{a_{1_U}, ..., a_{k_U}\}$
    - $D_{\text{with } f \text{ in } se} = \{x \in dom(f_U) \mid f_U(x) \in D_{se}\}$
    - $D_{\text{with } r \text{ in } se} = \{x \in dom(r_U) \mid (x,y) \in r_U \Rightarrow y \in D_{se}\}$
    - $D_{\text{some } r} = \{x \in dom(r_U) \mid \text{there is a } y \text{ with } (x,y) \in r_U\}$
    - $D_{\text{agree } fp\ fp'} = \{x \in dom(fp_U) \cap dom(fp'_U) \mid fp_U(x) = fp'_U(x)\}$
    - $D_{\text{disagree } fp\ fp'} = \{x \in dom(fp_U) \cap dom(fp'_U) \mid fp_U(x) \neq fp'_U(x)\}$

    where

    - $dom(f_U) = D_s$ *for* $f \in F_{s \rightarrow se}$
    - $dom(fp_U) = D_s$ *for* $fp \in FP_{s \rightarrow se}$
    - $dom(r_U) = D_s$ *for* $r \in R_{s,se}$
    - $(f;p)_U(x) = p_U(f_U(x))$

- F_U *is a set of functions containing a total function* $f_U : D_s \longrightarrow D_{se}$ *for each feature* $f \in F_{s \rightarrow se}$

- R_U *is a set of relations containing a relation* $r_U \subseteq D_s \times D_{se}$ *for each role* $r \in R_{s,se}$
- *an element* $a_U \in D_s$ *for each atom* $a \in A_s$ *such that* $a \neq a'$ *implies* $a_U \neq a'_U$

The interpretation of the sorts, atoms, features, and roles meets the intuition behind these concepts as syntactic means for speaking about sets, objects within these sets, one-place functions, and two-place relations, respectively. These basic notions are used to define those sets within the universe which are described by arbitrary sort expressions.

An important notion in relation with sort expressions is that of subsumption.

Definition 2.4 *Let* Σ^{SORT} *be a sort signature and* se *and* se' *be sort expressions over* Σ^{SORT}. *se' Σ^{SORT}-subsumes se iff for any universe* $U = \langle D, F_U, R_U \rangle$ *for* Σ^{SORT} *we have* $D_{se} \subseteq D_{se'}$. *We then write* $se \ll_{\Sigma SORT} se'$.

For a sort signature Σ^{SORT} the subsumption relation $\ll_{\Sigma SORT}$ is reflexive and transitive relation on $SE(\Sigma^{SORT})$. $\ll_{\Sigma SORT}$ is not a partial ordering because antisymmetry does not hold as the following counter example shows:

    sort $\top$
        atoms a

    sort s
        features $f, g : [\, a \,]$
then we have

    with f **in** $[a] \ll_{\Sigma SORT}$ **with** g **in** $[a]$

and

    with g **in** $[a] \ll_{\Sigma SORT}$ **with** f **in** $[a]$

but the two sort expressions are not identic.
In order to obtain a partial order on the sort expressions, we apply a standard trick and factorize the set of sort expressions by the following equivalence relation:

$$se \equiv_{\Sigma SORT} se' \text{ iff } D_{se} = D_{se'} \text{ in any universe } U = \langle D, F_U, R_U \rangle$$

Then we obtain a standard result

Lemma 2.5 *Let* Σ^{SORT} *be a sort signature. Then* $SE_{/\equiv_{\Sigma SORT}}(\Sigma^{SORT})$ *is a lattice with respect to the relation*

$$[se] \ll_{\Sigma SORT} [se'] \text{ iff } se \ll_{\Sigma SORT} se'$$

We will be a bit sloppy in the sequel and use sort expressions in situations where we should rather be more precise and use equivalence classes of sort expressions.

2.3 Consistency Considerations

Since universes are semantic structures for sort signatures, we can ask the standard question: 'Does a universe exist for a sort signature?'. Technically speaking we introduce a first concept of consistency.

Definition 2.6 Let $\Sigma^{SORT} = \langle S, A, F, R \rangle$ be a sort signature.
Σ^{SORT} is consistent iff there is a universe U for Σ^{SORT}.

Considering the sort signature given in an ad hoc concrete syntax

> **sort** s
> > **features** $f : s \sqcap \neg s$
> > > **atoms** a

we observe that sort signatures may be inconsistent.
The inconsistency of the above sort signature arises because the feature f cannot be interpreted (as a function). This holds due to the fact that in any universe, D, has to contain at least one element which enforces any universe to interpret f as a function having a non-empty range which is impossible since the range of f has to be the empty set because of $s \sqcap \neg s \ll_{\Sigma^{SORT}} \bot$.

Let us close this section by a brief comparison between L_{LILOG}, ordinary order-sorted logic, and KL-ONE.

- L_{LILOG} *vs. order-sorted logic*
  The sort signatures occuring in order-sorted predicate logic consist only of the set of sorts, i. e., no atoms, features, and roles are part of the sort signature if we adapt the concept of sort signatures to the setting of order-sorted predicate logic. For a sort signature containing only sort names, a universe interpreting these sort names can always be found. This is an important difference compared to L_{LILOG} sort signatures.

- L_{LILOG} *vs. KL-ONE*
  In the KL-ONE family of languages, sorts, atoms, features and roles are only declared implicitly by mentioning them in sort expressions. Adapting the concept of sort signature to the setting of KL-ONE, we also have a set of sorts. The atoms are all attached to $\top$ and the source and the target sort of the features and roles are both $\top$. Under these conditions the existence of a universe for a KL-ONE sort signature can always be guaranteed: just take the set of terms formed over the atoms and features as the interpretation of the sort $\top$, all other sorts can be interpreted as the empty set, interpret the features as it is usual for term models, and interpret the roles by empty relations.

2.4 On The Existence of Universes

Knowing that sort signatures may be inconsistent in general, we are now going to establish criteria which assure the existence of a universe interpreting a sort signature.
As sort signatures are part of the signature of a logical theory and term models are of particular interest, we are going to develop criteria for sort signatures which are strong enough to admit universes whose carrier sets are terms formed over the atoms and features of a sort signature.

The first step towards constructing a universe of terms for a sort signature is thus to define the family of terms that can be formed over a sort signature Σ^{SORT}.
The construction of the family of terms cannot be performed along the standard lines of order-sorted

logic since that would lead to too few terms. The main difference is that we have to impose quite a number of *additional closure conditions* due to the fact that sort expressions impose structural requirements on how to assign sets to them. And since terms are syntactic objects to be interpreted by elements of the sets interpreting sort expressions, one has to have enough terms of the respective sort expressions.

Definition 2.7 *Let* $\Sigma^{SORT} = \langle S, A, F, R \rangle$ *be a sort signature. The terms over* Σ^{SORT} *form the least* $SE(\Sigma^{SORT})$*-indexed family of sets*
$$T^{\Sigma^{SORT}} = \langle T_{se}^{\Sigma^{SORT}} \rangle_{se \in SE(\Sigma^{SORT})}$$
satisfying

- $a \in A$, *implies* $a \in T_{,}^{\Sigma^{SORT}}$
- $f \in F_{s \to se}, t \in T_{se'}^{\Sigma^{SORT}}$ *implies* $f(t) \in T_{se}^{\Sigma^{SORT}}$ *for* $se' \ll_{\Sigma SORT} s$

and the following closure conditions

- $t \in T_{se}^{\Sigma^{SORT}}$ *implies* $t \in T_{se'}^{\Sigma^{SORT}}$, *for* $se \ll_{\Sigma SORT} se'$
- $t \in T_{se}^{\Sigma^{SORT}} \cap T_{se'}^{\Sigma^{SORT}}$ *implies* $t \in T_{se \cap se'}^{\Sigma^{SORT}}$
- $t \in T_{se \cup se'}^{\Sigma^{SORT}}$ *implies* $t \in T_{se}^{\Sigma^{SORT}} \cup T_{se'}^{\Sigma^{SORT}}$
- $t \in T_{se}^{\Sigma^{SORT}}$ *iff* $t \notin T_{\neg se}^{\Sigma^{SORT}}$
- $a \in A_{,}$, *implies* $a \in T_{[a]}^{\Sigma^{SORT}}$
- $f \in F_{s \to \text{with } g \text{ in } se}, g \in F_{s' \to se'}, t \in T_{s}^{\Sigma^{SORT}}$ *implies* $g(f(t)) \in T_{se}^{\Sigma^{SORT}}$
- $f \in F_{s \to se}, t \in T_{s}^{\Sigma^{SORT}}$ *implies* $t \in T_{\text{with } f \text{-in } se}^{\Sigma^{SORT}}$
- $fp \in FP_{s \to se}, t \in T_{s}^{\Sigma^{SORT}}$ *implies* $t \in T_{\text{agree } fp \, fp}^{\Sigma^{SORT}}$

This rather involved definition of the family of terms is due to the structure of the sort expressions. We don't give the rationale for all of the items in the above definition but rather pick one of them and show why the standard mechanism for constructing terms does not generate all the terms one would like to have.

- $f \in F_{s \to se}, t \in T_{s}^{\Sigma^{SORT}}$ *implies* $t \in T_{\text{with } f \text{ in } se}^{\Sigma^{SORT}}$

This closure condition of the term definition assures that a term $t \in T_{s}^{\Sigma^{SORT}}$ is also contained in $T_{\text{with } f \text{ in } se}^{\Sigma^{SORT}}$ and not only in $T_{s}^{\Sigma^{SORT}}$. This should intuitively hold because the feature f applied to the object t gives an object of the sort se: thus we should have t being an element of $T_{\text{with } f \text{ in } se}^{\Sigma^{SORT}}$.

With the family of terms constructed according to the above definition we are now going to provide a first theorem on the existence of term universes for sort signatures. Due to the fact that arbitrary sort signatures may be inconsistent the theorem states criteria assuring the existence of a universe.

Theorem 2.8 *Let* $\Sigma^{SORT} = \langle S, A, F, R \rangle$ *be a sort signature such that*

1. $F_{s \to se} = \emptyset$, *if* $T_s \neq \emptyset$ *and* $se \ll_{\Sigma SORT} \perp$
2. $A_\perp = \emptyset$
3. $F_{s \to \text{with } g \text{ in } se} = \emptyset$ *for* $g \in F_{s' \to se'}$ *if* $T_s \neq \emptyset$ *and* $se \ll_{\Sigma SORT} \perp$
4. $F_{s \to se} \cap F_{s' \to se} = \emptyset$, *if* $T_s \cap T_{s'} \neq \emptyset$
5. $F_{s \to [a_1, ..., a_n]} = \emptyset$ *for any non-empty set of atoms* $\{a_1, ..., a_n\}$, *if* $T_s \neq \emptyset$

6. $T_s \subseteq T_{\text{with } r \text{ in } se'}$ *for* $r \in R_{s,se'}$, *if* $T_s \neq \emptyset$

7. $F_{s \rightarrow \text{some } r} = \emptyset$ *for* $r \in R_{s',se'}$, *if* $T_s \neq \emptyset$

8. $F_{s \rightarrow \text{agree } fp\ fp'} = \emptyset$ *for* $fp \in FP_{s_1,se_1}$, $fp' \in FP_{s_3,se_3}$ *and* $fp \neq fp'$, *if* $T_s \neq \emptyset$

9. $F_{s \rightarrow \text{disagree } fp\ fp} = \emptyset$ *for* $fp \in FP_{s',se'}$, *if* $T_s \neq \emptyset$

Then $T = \langle T^{\Sigma^{SORT}}, F_T, R_T \rangle$ *such that*

- $T^{\Sigma^{SORT}}$ *is the family of terms over* Σ^{SORT}

- *for any atom* $a \in A$, *we let* $a_T = a$

- $F_T = \langle F_{T,s \rightarrow se} \rangle_{s \in S, se \in SB(\Sigma^{SORT})}$
  is a family of functions containing for each feature $f \in F_{s \rightarrow se}$ *a function*
  $$f_T : T_s^{\Sigma^{SORT}} \longrightarrow T_{se}^{\Sigma^{SORT}}$$
  with
  $$f_T(t) = f(t)$$

- $R_T = \langle R_{T,s,se} \rangle_{s \in S, se \in SB(\Sigma^{SORT})}$
  is a family of relations containing for each role $r \in R_{s,se}$ *a relation*
  $$r_T \subseteq T_s^{\Sigma^{SORT}} \times T_{se}^{\Sigma^{SORT}}$$
  with
  $$r_T = \emptyset$$

is a universe for Σ^{SORT}.

The proof of the theorem can be found in [Pletat 90].
The universe for a sort signature Σ^{SORT} whose existence is assured through the above theorem is the simplest universe one can think of. It plays the role similar to a Herbrand universe in the sense that we are only interested in the domains and functions while the roles are interpreted as simple as possible: the relations associated to them are all empty.
As can be seen from the list of conditions of Theorem 2.8, a sort signature has to satisfy quite a number of criteria in order to obtain a universe for it. These conditions can be split into three classes:

- Conditions that must always be satisfied: this holds for conditions 1, 2, 3, 4, and 9.
  I.e., these conditions are also necessary ones for the existence of a model.

- Conditions that can be dropped if universes where the domains are not sets of terms but sets of equivalence classes of terms are considered: this holds for conditions 5 and 8.

- Conditions that can be dropped if universes with non-empty relations interpreting the roles are considered: this holds for conditions 6 and 7.

In the following we will sketch how to overcome the restrictions 5 and 8 of the theorem assuring the existence of universes. This can be done by means of factorization. The phenomenon common to both the restrictions imposed in conditions 5 and 8 is that they prevent the sets of terms interpreting the sort expressions $[a_1, ..., a_n]$ and **agree** $fp\ fp'$ to be - so to speak - polluted due to the fact that there are features having these sort expressions as their target while terms of their source sorts do exist. This pollution leads to too many non-identic terms in the sets assigned to the corresponding sort expressions. The factorization we are going to apply has the nice property of cleaning up the polluted sets of terms in T by forming equivalence classes through identification of terms.

Overcoming restriction 5
Assume the situation

$$F_{s \to [a_1,...,a_n]} \neq \emptyset \text{ for a non-empty set of atoms } \{a_1,...,a_n\} \text{ and } T_s^{\Sigma^{SORT}} \neq \emptyset$$

to be given.
For a fixed $a \in \{a_1,...,a_n\}$ we define the following set of equations

$$E(a) = \{f(t) \doteq a \mid f \in F_{s \to [a_1,...,a_n]}, t \in T_s^{\Sigma^{SORT}}\}.$$

Then $E(a)$ generates a least congruence $\equiv_{E(a)}$ on T, see [Ehrig, Mahr 85].

- Taking the quotient
$$T_{/\equiv_{E(a)}}^{\Sigma^{SORT}} \text{ of } T^{\Sigma^{SORT}}$$
as the domain

- and defining functions
$$f_{T_{/\equiv_{E(a)}}} : (T_{/\equiv_{E(a)}}^{\Sigma^{SORT}})_s \longrightarrow (T_{/\equiv_{E(a)}}^{\Sigma^{SORT}})_{se}$$
by
$$f_{T_{/\equiv_{E(a)}}}([t]) = [f(t)]$$

we obtain a universe $T_{/\equiv_{E(a)}}$ for Σ^{SORT} (where the roles are still interpreted as empty sets) which satisfies the condition

$$D_{[a_1,...,a_n]} = \{a_{1_U},...,a_{n_U}\}$$

for interpreting enumeration sort expressions. This holds due to the fact that any term $f(t) \in T_{[a_1,...,a_n]}^{\Sigma^{SORT}}$ is identified with the atom a while the atoms are still interpreted by pairwise different elements of the universe.
Taking into account that we have chosen an arbitrary atom a from the set of atoms $\{a_1,...,a_n\}$ in the factorization for overcoming restriction 5, we observe that universes for sort signatures not satisfying condition 5 are in general not uniquely determined in situations where we have $n \geq 2$.

Overcoming restriction 8
Assume the situation

$$F_{s \to \text{agree } fp\, fp'} \neq \emptyset \text{ for } fp \in FP_{s_1,se_1} \text{ and } fp' \in FP_{s_2,se_2} \text{ with } fp \neq fp', \text{ and } T_s^{\Sigma^{SORT}} \neq \emptyset$$

to be given.
We define the following set of equations

$$E = \{fp(f(t)) \doteq fp'(f(t)) \mid f \in F_{s \to \text{agree } fp\, fp'}, t \in T_s^{\Sigma^{SORT}}\}$$

Then E generates a least congruence $\equiv_E$ on $T^{\Sigma^{SORT}}$.

- Taking the quotient
$$T_{/\equiv_E}^{\Sigma^{SORT}} \text{ of } T^{\Sigma^{SORT}}$$
as the domain

- and defining functions

$$f_{\mathcal{T}/_{\equiv_{\mathbf{s}}}} : (T_{/_{\equiv_{\mathbf{s}}}}^{\Sigma^{SORT}})_{\cdot} \longrightarrow (T_{/_{\equiv_{\mathbf{s}}}}^{\Sigma^{SORT}})_{\cdot e}$$

by

$$f_{\mathcal{T}/_{\mathbf{s}_{\mathbf{s}}}}([t]) = [f(t)]$$

we obtain a universe $\mathcal{T}/_{\equiv_{\mathbf{s}}}$ for Σ^{SORT} (where the roles are still interpreted as empty sets) which satisfies the condition

$$fp([z]) = fp'([z]) \text{ for all } [z] \in (T_{/_{\equiv_{\mathbf{s}}}}^{\Sigma^{SORT}})_{\text{agree } fp \, fp'}$$

This holds due to the fact that for any term $f(t) \in T_{\text{agree } fp \, fp'}^{\Sigma^{SORT}}$ the identification of $fp(f(t))$ and $fp'(f(t))$ has been enforced by the factorization.

The previous discussion shows that universes do exist for a wider class of sort signatures than the one specified in the premise of Theorem 2.8. An observation we can make is that for this wider class of sort signatures we do not have universes whose domains are just sets of terms. Instead, we have to consider quotients of term universes because of the factorizations that became necessary in order to establish the existence of universes under the relaxed preconditions.

The next step is to show that also the restrictions 6 and 7 of Theorem 2.8 can be overcome. The 'price' one has to pay for this is that universes whose relations are non-empty have to be admitted.

Overcoming restriction 6

Assume the situation

$$T_{\cdot}^{\Sigma^{SORT}} \not\subseteq T_{\text{with } r \text{ in } \cdot e'}^{\Sigma^{SORT}} \text{ for } r \in R_{\cdot, \cdot e} \text{ and } T_{\cdot}^{\Sigma^{SORT}} \neq \emptyset$$

to be given. If this is the case, we have a term $t \in T_{\cdot}^{\Sigma^{SORT}} \setminus T_{\text{with } r \text{ in } \cdot e'}^{\Sigma^{SORT}}$.
In order to assure

$$t \notin \{z \in T_{\cdot}^{\Sigma^{SORT}} \mid (z, z') \in r_{\mathcal{T}} \Rightarrow z' \in T_{\cdot e'}^{\Sigma^{SORT}}\}$$

we have

1. to avoid the existence of a pair $(t, t') \in r_{\mathcal{T}}$ for $t' \in T_{\cdot e'}^{\Sigma^{SORT}}$

2. to be able to establish a pair $(t, t') \in r_{\mathcal{T}}$ for $t' \in T_{\cdot e}^{\Sigma^{SORT}} \setminus T_{\cdot e'}^{\Sigma^{SORT}}$.

Under the assumption that $T_{\cdot e}^{\Sigma^{SORT}} \setminus T_{\cdot e'}^{\Sigma^{SORT}} \neq \emptyset$ the following construction of $r_{\mathcal{T}}$ leads to a universe for a sort signature not satisfying restriction 6 of Theorem 2.8:

- $r_{\mathcal{T}} \subseteq [T_{\cdot}^{\Sigma^{SORT}} \setminus T_{\text{with } r \text{ in } \cdot e'}^{\Sigma^{SORT}}] \times [T_{\cdot e}^{\Sigma^{SORT}} \setminus T_{\cdot e'}^{\Sigma^{SORT}}] \cup [T_{\text{with } r \text{ in } \cdot e'}^{\Sigma^{SORT}}] \times [T_{\cdot e}^{\Sigma^{SORT}} \cap T_{\cdot e'}^{\Sigma^{SORT}}]$
  with
  $[T_{\text{with } r \text{ in } \cdot e'}^{\Sigma^{SORT}}] \times [T_{\cdot e}^{\Sigma^{SORT}} \cap T_{\cdot e'}^{\Sigma^{SORT}}] = \emptyset$

- for any $t \in T_{\cdot}^{\Sigma^{SORT}} \setminus T_{\text{with } r \text{ in } \cdot e}^{\Sigma^{SORT}}$ we let the pair $(t, t') \in r_{\mathcal{T}}$ where t' is a fixed element of $T_{\cdot e}^{\Sigma^{SORT}} \setminus T_{\cdot e'}^{\Sigma^{SORT}}$.

The above construction exhibits a further hint on what kinds of universes can exist for a sort signature not satisfying restriction 5 of Theorem 2.8. The requirement that $T_{se}^{\Sigma^{SORT}} \setminus T_{se'}^{\Sigma^{SORT}}$ must not be empty is not essential for the existence of any universe, but only in the situation where we are focusing on term universes. If $T_{se}^{\Sigma^{SORT}} \setminus T_{se'}^{\Sigma^{SORT}} = \emptyset$ holds, we can at most expect universes for a sort signature that are not term-generated, see [Goguen, Burstall 85].

As long as one is not concerned with operational aspects of languages of the kind we are discussing here this does no harm. However, when implementing an inference component for L_{LILOG} we would have to insist on the requirement $T_{se}^{\Sigma^{SORT}} \setminus T_{se'}^{\Sigma^{SORT}} \neq \emptyset$ because the model theory underlying typical theorem provers is restricted to term-generated models, see [Robinson 65].

Finally we want to outline how the condition 7 of Theorem 2.8 can be overcome if universes whose relations are non-empty are taken into account.

Overcoming restriction 7

Assume the situation

$$F_{s \rightarrow some \, r} \neq \emptyset \ for \ r \in R_{s',se'} \ and \ T_s^{\Sigma^{SORT}} \neq \emptyset$$

to be given. Then we have $f(t) \in T_{some \, r}^{\Sigma^{SORT}}$ for some term $t \in T_s^{\Sigma^{SORT}}$. Thus, in order to satisfy the requirement for interpreting the sort expression some r, we have to have a term $t' \in T_{se'}^{\Sigma^{SORT}}$ and $(f(t), t')$ should be an element of r_T.

Under the assumption that $T_{se'}^{\Sigma^{SORT}} \neq \emptyset$ the following construction for r_T leads to a universe for a sort signature not satisfying condition 7 of Theorem 2.8.

- Let t' be a fixed element of $T_{se'}^{\Sigma^{SORT}}$. Then we define

$$r_T = \{(f(t), t') \mid f \in F_{s \rightarrow some \, r}, t \in T_s^{\Sigma^{SORT}}\}$$

As in the construction overcoming restriction 6, the requirement $T_{se'}^{\Sigma^{SORT}} \neq \emptyset$ is only essential for the existence of term-generated universes. If one is willing to accept any kind of universe for a sort signature, it can be dropped. Then, however, it may be the case that one can at most have universes for a sort signature that are not term-generated.

3 Sort Hierarchies

In the previous section we have seen that sort signatures can be inconsistent, i.e., there are sort signatures for which we cannot find a universe.

The idea behind the sort hierarchies to be introduced next is to influence the subsumption ordering $\ll_{\Sigma^{SORT}}$ on sort expressions by imposing so-called sort constraints. These constraints allow us to require a sort to be interpreted by the same set as a sort expression, i.e., we may enforce certain identities between the sets of a universe; and these identities should then hold in any universe for a sort signature.

Imposing such sort constraints for sort signatures which are inconsistent doesn't make much sense since these sort signatures don't have a universe and thus will satisfy any sort constraint. Therefore we make the following

General Assumption: *Any sort signature has to be consistent.*

Definition 3.1 *Let* $\Sigma^{SORT} = \langle S, F, R, A \rangle$ *be a sort signature. The set of sort constraints* $SC(\Sigma^{SORT})$ *over* Σ^{SORT} *is defined as*

- $SC(\Sigma^{SORT}) = \{s \doteq se \mid s \in S, se \in SE(\Sigma^{SORT})\}$

Using sort constraints of the form $s \doteq se$ we can define other constraints which will be of interest. For example we can express the constraint $s \leq se$ by $\neg se \sqcap s \doteq \bot$, or $s \doteq s \sqcap se$, or $s \doteq se \sqcap s'$ for a new sort name s'. The disjointness condition for two sorts can be stated directly as $s \sqcap s' = \bot$.

Definition 3.2 *A sort hierarchy* $SH = \langle \Sigma^{SORT}, SC \rangle$ *consists of a sort signature* Σ^{SORT} *and a set of sort constraints* SC *over* Σ^{SORT}.

Sort hierarchies are a generalized form of the partially ordered set of sorts which usually describe the set structure of the universe of discourse in knowledge bases defined in order-sorted logic, see [Walther 87]. Looking at the KL-ONE world, sort hierarchies correspond to (restricted) T-Boxes, c.f. [Brachman et al. 83].

Sort hierarchies impose constraints on the interpretation of sort names by enforcing that a sort name has to be interpreted by a special set, namely the set defined by the sort expression forming the righthand side of the sort constraint.

Returning to our example on vehicles, we define *sportscars* as vehicles having 4 wheels, 2 doors and which are of type *cabrio*. This can be done by making the following sort constraint part of the sort hierarchy:

> **sort** *sportscar* = **with-feature** *wheels* **in** [4] $\sqcap$
>   **with-feature** *doors* **in** [2] $\sqcap$
>   **with-feature** *type* **in** [*cabrio*]

Now we want to define the meaning of a sort constraint and introduce the concept of a universe satisfying the constraints of a sort hierarchy. This is straightforward and the definition below is what we expect intuitively.

Definition 3.3 *Let* $\Sigma^{SORT} = \langle S, F, R, A \rangle$ *be a sort signature and* $U = \langle D, F_U, R_U \rangle$ *a universe for* Σ^{SORT}. *U satisfies the sort constraint*

- $s \doteq se$ *iff* $D_s = D_{se}$; *we then write* $U \models s \doteq se$.

All these concepts are put together in order to define whether a universe is a semantic structure satisfying the sort constraints of a sort hierarchy.

Definition 3.4 *Let* $SH = \langle \Sigma^{SORT}, SC \rangle$ *be a sort hierarchy and* $U = \langle D, F_U, R_U \rangle$ *a universe for* Σ^{SORT}.
U is a universe for SH iff U is a universe Σ^{SORT} *and* $U \models sc$ *for any* $sc \in SC$.

Because of

$$U \models s \doteq se \text{ and } U \models s \doteq se' \text{ iff } U \models s \doteq se \sqcap se'$$

we may assume that in any sort hierarchy for any sort s we have at most one sort constraint where s forms the lefthand side.
Because of

$$U \models s_1 \doteq s \text{ and } U \models s_2 \doteq s \text{ iff } U \models s \doteq s_1 \sqcap s_2$$

we may assume that in any sort hierarchy for any sort s we have only one sort constraint where s forms the righthand side.

The sort constraints can be considered as axioms formulated over the symbols declared in the sort signature. These axioms are expressive enough to make sort hierarchies inconsistent, i. e. there are sort hierarchies for which no universe exists. The following sort hierarchy is an example of this kind:

sort $s \doteq [a]$
    atoms a, b

sort s'

When introducing the language construct of sort hierarchies new sources for inconsistency shape up. At the moment we are not able to make as detailed descriptions about the existence of models as in Section 2.

The inconsistency of sort hierarchies is again in contrast to what we know from order-sorted logic and the KL-ONE world.

- In order-sorted predicate logic we may have only sort constraints of the form

$$s \doteq s \sqcap s' \text{ or } s' \doteq s' \sqcup s$$

expressing

$$s \leq s'.$$

If the sort signature consists only of sort names and we have only sort constraints requiring $s \leq s'$ in our hierarchy, the hierarchy is consistent. The worst thing that can happen is to require $s \leq \bot$ which forces s to be interpreted as the empty set. This, however, cannot cause inconsitencies.

- Considering the setting given in the KL-ONE context, we recall that there the sort signatures consisted of a set of sort names, plus atoms, features, and roles attached solely to the top sort $\top$. As long as cycle-free sort hierarchies are considered, universes can always be found for KL-ONE T-Boxes since constraints that require sorts to be interpreted by non-empty sets are typically forbidden. Thus a universe interpreting all sorts (except for the top sort $\top$) by the empty set can always be chosen. For restricted forms of T-Boxes containing cyclic definitions of sorts by sort constraints, [Nebel 89] shows the existence of universe for any sort signature. It is unclear to us whether the results of [Nebel 89] carry over to our setting.

On the basis of the concept of a universe for a sort hierarchy we can define a second notion of subsumption for sort expressions: subsumption with respect to a sort hierarchy SH.

$$se \ll_{SH} se' \text{ iff } D_{se} \subseteq D_{se'}, \text{ for all universes } U = \langle D, F_U, R_U \rangle \text{ for } SH.$$

According to this definition we obtain $s \ll_{SH} s'$ for our sort hierarchy above, because it is inconsistent. The sort signature alone is still consistent, but we do not obtain $s \ll_{\Sigma SORT} s'$. The following lemma is an immediate consequence of the above definition

Lemma 3.5 Let $SH = \langle \Sigma^{SORT}, SC \rangle$ be a sort hierarchy. Then we have for any two sort expressions se and se': $se \ll_{\Sigma SORT} se'$ implies $se \ll_{SH} se'$.

4 Signatures

The previous two sections introduced the syntactic and semantic concepts for describing the universe of discourse of a L_{LILOG} knowledge base in terms of a sort hierarchy. What we have seen there was a generalisation of the usual concept of a partially ordered set of sorts, which is used in ordinary order-sorted logic to define sort hierarchies, in order to integrate some ideas known from KL-ONE and feature term languages.

The next step is to introduce signatures containing the operator symbols and predicate symbols over which terms and formulas can be formed in order to provide the axioms of a knowledge base.

Being guided by order-sorted logic where the sort hierarchies of the signatures are always consistent we make the

> **General Assumption:** *Any sort hierarchy is consistent.*

Definition 4.1 *A signature* $\Sigma = \langle SH, O, P \rangle$ *consists of*

- *a sort hierarchy* $SH = \langle \Sigma^{SORT}, SC \rangle$
- *a family of sets of operators* $O = \langle O_{w \to se} \rangle_{w \in SB(\Sigma^{SORT})^*, se \in SB(\Sigma^{SORT})}$ *over* Σ^{SORT}
- *a family of sets of predicates* $P = \langle P_w \rangle_{w \in SB(\Sigma^{SORT})^*}$ *over* Σ^{SORT}

Signatures define the alphabet of symbols over which terms and formulas shall be constructed. We do already know about the sort hierarchy being part of a signature and we will discuss the next components below.

The operators are those function symbols which we use for the term construction. The predicates play the standard role as in any logic. Features, atoms and roles of a sort signature are also operations and predicates, respectively, of a signature.

Definition 4.2 *Let* $\Sigma^{SORT} = \langle S, A, F, R \rangle$ *be a sort signature.*

- *A family of operators* O *over* Σ^{SORT} *is a family of sets*
  $$O = \langle O_{w \to se} \rangle_{w \in SB(\Sigma^{SORT})^*, se \in SB(\Sigma^{SORT})}$$
  such that

  - $A_s \subseteq O_{\epsilon \to s}$, *for* $s \in S$, *i.e. atoms may appear as constants.*
    (*ϵ stands for the empty string*)
  - $F_{s \to se} \subseteq O_{s \to se}$, *for* $s \in S, se \in SE(\Sigma^{SORT})$, *i.e. features may be used as operators.*
- *A family of predicates* P *over* Σ^{SORT} *is a family of sets*
  $$P = \langle P_w \rangle_{w \in SB(\Sigma^{SORT})^*}$$
  such that

  - $R_{s,se} \subseteq P_{s,se}$ *for any* $s \in S, se \in SE(\Sigma^{SORT})$, *i.e. roles appear as binary predicates.*

The idea of signatures is to introduce further constants, function and predicate symbols for forming terms and logical axioms. While atoms, features, and roles are symbols of a signature which may be used for defining sorts this is not possible for the other symbols of the signature.

We continue our story on traveling people and discuss an important predicate that should be part of any signature of a knowledge base on traveling:

> **predicate** *travel* : *person*, *location*, *location*, *vehicle*

Having the above predicate in the signature requires us to introduce the sorts *person* and *person* by

> **sort** *person*

and

> **sort** *person* < *person*

where *person* stands for the sort of finite subsets of *person* plus the base sort *person*, see also [Pletat, Luck 89].

Although we didn't speak about literals up to now, it should be clear what they look like; thus with the above predicate being part of a signature we may express that 'John and Mary travel with their Porsche from LA to SF' by the literal *travel(John ⊕ Mary, LA, SF, Porsche)*

assuming the operator

operator ⊕ : *person, person → person*

to be contained in the signature. In addition, the persons John and Mary, the cities LA and SF, and the Porsche have to be introduced as objects of the knowledge base and should thus be mentioned in its signature; they would typically be introduced as constants:

constant *John : person*

constant *Mary : person*

constant *LA : location*

constant *SF : location*

constant *Porsche : sportscar*

The syntactic concept of a signature shall now be given a meaning by introducing the notion of a model for a signature Σ.

Definition 4.3 *Let $\Sigma = \langle SH, O, P \rangle$ be a signature where $SH = \langle \Sigma^{SORT}, SC \rangle$ with $\Sigma^{SORT} = \langle S, A, F, R \rangle$.*
A Σ-__model__ is a triple $M = \langle U, F_M, R_M \rangle$ where

- *U is a universe for SH*
- *F_M is a set of functions containing for each operator $o \in O_{w \to se}$ a function $o_M : D_w \longrightarrow D_{se}$ such that $o_M = o_U$, if o is a feature or an atom.*
- *R_M is a set of relations containing for each predicate $p \in P_w$ a relation $p_M \subseteq U_w$ such that $p_M = p_U$ if p is a role.*

Apart from the involved definition of the universe of a model we have given before, this notion is nothing but the straightforward extension of classical notions of models for a signature to our setting where sorts are not just names but complex expressions.

Again we face the problem that models need not exist for any L_{LILOG} signature which is in contrast to ordinary order-sorted logic.
The sources for inconsistency seem to remain conceptually the same as in Section 2 (disregarding the reasons for inconsistency of sort hierarcies!). This means that we will have to impose basically the same restrictions on the family of operators of a signature as we have imposed on the families of atoms and features when considering only sort signatures in Theorem 2.8 in order to assure the existence of a term model for a signature. We must, however, generalize to the situation that the source of an operator symbol may be a string of sort expressions rather than only a sort name.

In order-sorted predicate logic an inconsistent signature can only occur if the sort constraint $s \leq \bot$ is part of the sort hierarchy and s is the target sort of an operator $o : w \longrightarrow s$ for which we are not enforced to interpret w as the empty set. Approaches to order-sorted logic typically forbid the explicit use of the bottom sort $\bot$ (see, e.g., [Walther 87]) and consider it only as an additional symbol used for technical reasons. Forbidding the use of the bottom sort in this way within typical approaches to order-sorted logic leads to the well-known situation that consistency problems with the signatures arising for our logic are not present in order-sorted logic.

Note that the signatures of KL-ONE A-boxes do not allow for further operators and predicates than the atoms, features, and roles, respectively, introduced in the sort signatures. These additional symbols may, however, become part of the signature in KRYPTON knowledge bases.

Since the models of a signature $\Sigma = \langle SH, O, P \rangle$ may require different universes than the sort hierarchy SH alone, a third notion of subsumption for sort expressions has to be defined.

Let Σ be a signature and se and se' be two sort expressions.

$se \ll_\Sigma se'$ iff for any Σ-model $M = \langle U, F_M, R_M \rangle$ we have

$D_{se} \subseteq D_{se'}$ assuming $U = \langle D, F_U, R_U \rangle$.

We then say that se' Σ-subsumes se.

5 Terms

The next step in the definition of L_{LILOG} is to define the construction of terms.

As terms shall contain variables let us first say what they are.

Definition 5.1 Let $\Sigma^{SORT} = \langle S, A, F, R \rangle$ be a sort signature. A family of variables V over Σ^{SORT} is a family of sets $V = \langle V_{se} \rangle_{se \in SB(\Sigma^{SORT})}$.

In Definition 2.7 we have already shown how to construct terms over a sort signature Σ^{SORT}. Given a signature Σ and a family of variables V we construct the family of terms $T^\Sigma(V)$ over Σ and V in complete analogy with three generalizations:

- Variables are terms

- Instead of features we have to consider arbitrary operators

- Any reference to the the subsumption relation $\ll_{\Sigma^{SORT}}$ has to be replaced by the new subsumption relationship $\ll_\Sigma$.

In Section 2 we did not comment the semantic condition

$$se' \ll_{\Sigma^{SORT}} s$$

used in the definition of terms. Let us briefly discuss what this is about in the general setting of constructing terms over a signature where $se' \ll_{\Sigma^{SORT}} s$ becomes $se' \ll_{\Sigma^{SORT}} se$ due to the fact that an operator may have arbitrary sort expressions as arguments not only a sort name.

Why do we have this *semantic* restriction in the *syntactic* concept of term construction in our logic? Let's have a look at what happens due to this condition.

The condition assures that $o(t)$, for example, can only be a term if for the term t of sort se' we can assure that in any model we have $D_{se'} \subseteq D_{se}$, if se is the source sort of the operator o.

This is nothing but a standard restriction for the term construction we find in other logics as well. However, there this condition can easily be formulated by purely syntactic means:

- *One-sorted predicate logic*
  Since we have only one sort name our semantic condition $se' \ll_\Sigma se$ is met trivially and no restrictions have to be imposed explicitly.

- *Many-sorted predicate logic*
  Since we may have only several sort names our semantic condition $se' \ll_\Sigma se$ is met by the well-known syntactic condition $se = se'$.

- *Order-sorted predicate logic*

  We may have only several sort names together with a partial ordering $<$ between them. In this situation our semantic condition $se' \ll_\Sigma se$ is satisfied if we have $se' \leq se$ where $\leq$ is the reflexive and transitive closure of $<$. Again, this can be defined easily on the level of syntax.

We use the semantic restriction on the sorts for our term construction because syntactic criteria assuring $se' \ll_\Sigma se$ are much harder to define for our complex sorts. Of course, one should formulate syntactic conditions for the sort expressions as part of the definition of term construction assuring the semantic property one has to impose. For a sort description language being as rich as ours, these syntactic conditions are at least as involved as testing the validity of propositional logic formulas. Currently we are not sure whether the property $se' \ll_\Sigma se$ is decidable at all, a phenomenon that may occur for rich sort description laguages, see [Schmidt-Schauss 88].

Before we consider the semantics of terms let's have a look on the influence of inconsistent signatures on the term construction.

Since our semantic condition $se' \ll_\Sigma se$ always holds for inconsistent signatures, the term construction will then generate any term that could be obtained for an unsorted logic. I. e. $T^\Sigma(V)$ would contain terms that are *syntactically ill-typed* in the sense that the term t may be of a sort which is greater than the source sort of the operator o when forming $o(t)$. But since inconsistent signatures don't have any model at all, forming ill-typed terms doesn't make much sense since they cannot be evaluated anyway.

The conclusion we draw from these considerations is:

Term construction makes only sense for consistent signatures!

Again, in logics offering a weaker type system such an assumption about the consistency of the signature over which terms shall be formed is not stated explicitly since the corresponding signatures are typically consistent.

The term evaluation can be defined along the standard lines.

Definition 5.2 *Let* $\Sigma = \langle SH, O, P \rangle$ *be a signature where* $SH = \langle \Sigma^{SORT}, SC \rangle$ *with* $\Sigma^{SORT} = \langle S, A, F, R \rangle$ *and* V *a family of variables over* Σ^{SORT}. *Let* $M = \langle U, O_M, P_M \rangle$ *be a model for* Σ.

- *A variable assignment* α *is a family of total functions*
  $$\alpha = \langle \alpha_{se} : V_{se} \longrightarrow D_{se} \rangle_{se \in SE(\Sigma^{SORT})}.$$

- *A variable assignment* α *induces a term evaluation*
  $$\bar{\alpha} = \langle \bar{\alpha}_{se} : T_\Sigma(V)_{se} \longrightarrow D_{se} \rangle_{se \in SE(\Sigma^{SORT})}$$
  where
  $$\bar{\alpha}(t) = \begin{cases} o_M & , if\ o \in O_{\epsilon \to se} \\ \alpha(v) & , if\ v \in V_{se} \\ o_M(\bar{\alpha}(t_1), ..., \bar{\alpha}(t_k)) & , if\ t = o(t_1, ..., t_k)\ and\ o \in O_{w \to se} \end{cases}$$

6 Formulas and Knowledge Bases

The final step in the definition of L_{LILOG} is to introduce formulas and knowledge bases and to define their semantics.

The construction of formulas is (at the moment) a restricted version of the standard set of formulas of first order predicate logic: we allow only clauses written in the form of rules as axioms of a knowledge base.

Definition 6.1 *Let* $\Sigma = \langle SH, O, P \rangle$ *be a* L_{LILOG} *signature where* $SH = \langle \Sigma^{SORT}, SC \rangle$ *and* V *a family of variables over* Σ^{SORT}.

- *The literals over* Σ *and* V *form the set*
  $$L^{\Sigma}(V) = \{p(t_1, ..., t_k), \neg p(t_1, ..., t_k) \mid p \in P_{se_1, ..., se_k} \text{ and } t_i \in T^{\Sigma}(V)_{se'_i} \text{ for } se'_i \ll_{\Sigma} se_i\}$$

- *The conjunctions over* Σ *and* V *form the set*
  $$\overline{CON^{\Sigma}}(V) = \{l_1 \wedge ... \wedge l_k \mid l_i \in L^{\Sigma}(V)\}$$

- *The disjunctions over* Σ *and* V *form the set*
  $$\overline{DIS^{\Sigma}}(V) = \{l_1 \vee ... \vee l_k \mid l_i \in L^{\Sigma}(V)\}$$

- *The formulas over* Σ *and* V *form the set*
  $$\overline{F^{\Sigma}}(V) = \{d \leftarrow c \mid d \in DIS^{\Sigma}(V), c \in CON^{\Sigma}(V)\}$$

For our predicate *travel* we may now introduce an axiom expressing that any member of a group of people traveling with some vehicle uses the same vehicle as the entire group.

> **forall** M : *person*, G : *person**, F : *location*, T : *location*, V : *vehicle* .
> $\qquad travel(M, F, T, V)$
> $\qquad\qquad \leftarrow travel(G, F, T, V), M \leq G$

With all these notions we can define what we want to consider as a L_{LILOG} knowledge base.

Definition 6.2 *A* L_{LILOG} *knowledge base* $KB = \langle \Sigma, AX \rangle$ *consists of*

- *a signature* Σ
- *a set of axioms* AX *such that an axiom* $ax \in AX$ *is an element of* $F^{\Sigma}(V)$ *where* V *is some family of variables* V *over* Σ^{SORT}, *the sort signature of* Σ.

For these last syntactic concepts of L_{LILOG} we now define their semantics in terms of the satisfaction relation between formulas and models and finally we introduce a loose semantics for L_{LILOG} knowledge bases.

Definition 6.3 *Let* $\Sigma = \langle SH, O, P \rangle$ *be a signature where* $SH = \langle \Sigma^{SORT}, SC \rangle$ *and* V *a family of variables over* Σ^{SORT}. *Let* $M = \langle U, O_M, P_M \rangle$ *with* $U = \langle D, F_U, R_U \rangle$ *be a* Σ-*model and* $\alpha : V \longrightarrow D$ *a variable assignment.*
M satisfies

- *the literal* $p(t_1, ..., t_k)$ *wrt* α *iff* $(\bar{\alpha}(t_1), ..., \bar{\alpha}(t_k)) \in p_M$; *we then write* $M \models_{\alpha} p(t_1, ..., t_k)$
- *the literal* $\neg p(t_1, ..., t_k)$ *wrt* α *iff* $(\bar{\alpha}(t_1), ..., \bar{\alpha}(t_k)) \notin p_M$; *we then write* $M \models_{\alpha} \neg p(t_1, ..., t_k)$
- *the conjunction* $l_1 \wedge ... \wedge l_k$ *wrt* α *iff* $M \models_{\alpha} l_i$ *for all i; we then write* $M \models_{\alpha} l_1 \wedge ... \wedge l_k$
- *the disjunction* $l_1 \vee ... \vee l_k$ *wrt* α *iff* $M \models_{\alpha} l_i$ *for some i; we then write* $M \models_{\alpha} l_1 \vee ... \vee l_k$
- *the formula* $d \leftarrow c$ *iff for all variable assignments* α *we have* $M \models_{\alpha} c$ *implies* $M \models_{\alpha} d$; *we then write* $M \models d \leftarrow c$.

Now we can define what is a model of a L_{LILOG} knowledge base.

Definition 6.4 *Let* $KB = \langle \Sigma, AX \rangle$ *be a* L_{LILOG} *knowledge base and* M *a model for* Σ. *We call* M *a model of* KB *iff* M *satisfies all the axioms in* AX.

We choose a loose semantics for a knowledge base according to

Definition 6.5 *Let* $KB = \langle \Sigma, AX \rangle$ *be a* L_{LILOG} *knowledge base. The semantics of KB is* **mod(KB)** *, the class of all models of KB.*

Consistency of knowledge bases becomes a standard notion.

Definition 6.6 *Let* $KB = \langle \Sigma, AX \rangle$ *be a* L_{LILOG} *knowledge base. KB is consistent iff there is a model for KB.*

As far as consistency of knowledge bases is concerned, we are back on the grounds of (e.g.) order-sorted predicate logic facing well-known problems.

Finally we give a last definition of subsumption of sort expressions:

$se \ll_{KB} se'$ iff $D_{se} \subseteq D_{se'}$ for any model $M = \langle U, F_M, R_M \rangle$ of KB where $U = \langle D, F_U, R_U \rangle$.

In anlogy to the class **mod(KB)** of all models of a knowledge base we define the classes of all universes of a sort signature Σ^{SORT} as **univ(Σ^{SORT})** , the class of all universes of a sort hierarchy SH to be **univ(SH)** and the class of all models of a signature Σ as **mod(Σ)** . Then we obtain:

- **univ**$(SH) \subseteq$ **univ**(Σ^{SORT}), for a sort hierarchy $SH = \langle \Sigma^{SORT}, SC \rangle$

- **mod**$(KB) \subseteq$ **mod**(Σ), for a knowledge base $KB = \langle \Sigma, V, AX \rangle$

Considering a knowledge base $KB = \langle \Sigma, AX \rangle$ with $\Sigma = \langle SH, O, P \rangle$ and $SH = \langle \Sigma^{SORT}, SC \rangle$, we obtain the following relationship of the various notions of subsumption for sort expressions due the above inclusions of the classes of universes and models, respectively, and due to the fact that a universe of a Σ-model is also a universe for a SH.

- $se \ll_{\Sigma^{SORT}} se'$ implies $se \ll_{SH} se'$ implies $se \ll_{\Sigma} se'$ implies $se \ll_{KB} se'$.

7 Conclusions

The stepwise definition of the syntax and semantics of a fragment of the knowledge representation language L_{LILOG} has exhibited several sources of inconsistency that may be present in L_{LILOG} knowledge bases. The problems of inconsistency we find in L_{LILOG} sort signatures, sort hierarchies, and signatures are not present in both parents of L_{LILOG} : order-sorted predicate logic and set description languages. The investigations of the phenomena reported in this paper have just begun and should lead to more results describing the existence of models for signatures in greater detail. What we aim at are syntactic criteria assuring the consistency of signatures that can be checked by knowledge engineering tools we would like to offer for the use of L_{LILOG} .

References

[Beierle et al. 89] C. Beierle, J. Dörre, U. Pletat, C. Rollinger, P.H. Schmitt, R. Studer: *The Knowledge Representation Language L_{LILOG}* . Proc. 2nd Workshop on Computer Science Logic 1988, E. Börger, H. Kleine Büning, M. M. Richter (eds), Lecture Notes in Computer Science, Vol. 385, Springer Verlag, Berlin 1989

[Beierle et al. 88] C. Beierle, U. Pletat, H. Uszkoreit: *An Algebraic Characterization of STUF* Proc. Symposium Computerlinguistik und ihre theoretischen Grundlagen, I.S. Batori, U. Hahn, M. Pinkal, W. Wahlster (eds), Informatik Fachberichte, Vol. 195, Springer Verlag, Berlin 1988

[Beierle et al. 89]   C. Beierle, U. Hedtstück, U. Pletat, J. Siekmann: *An Order Sorted Predicate Logic with Closely Coupled Taxonomic Information.* To appear, 1989

[Bouma et al. 88]   G. Bouma, E. König, H. Uszkoreit: *A flexible graph-unification formalism and its application to natural-language processing.* IBM J. Res. Develop. **32** 2, 1988, 170-184

[Brachman 79]   R.J. Brachman: *On The Epistomological Status of Semantic Networks* In: Associative Networks: Representation and Use of Knowledge By Computers, N. V. Findler (ed.), Academic Press, New York 1979

[Brachman et al. 83]   R.J. Brachman, R.E. Fikes, H.J. Levesque: *KRYPTON: Integrating Terminology and Assertion.* Proc. of AAAI-83 1983 31-35

[Brachman, Schmolze 85]   R.J. Brachman, J.G. Schmolze: *An Overview of the KL-ONE Knowledge Representation System.* Cognitive Science **9** (2) April 1985, 171-216

[Brachman et al. 85]   R.J. Brachman, V. Pigman, Gilbert, H.J. Levesque: *An Essential Hybrid Reasoning System.* Proc. IJCAI-85, 1985, 532-539

[Cohn 87]   A. G. Cohn: *A More Expressive Formulation of Many Sorted Logic.* Journal of Automated Reasoning, 3:113-200, 1987

[Ehrig, Mahr 85]   H. Ehrig, B. Mahr: *Foundations of Algebraic Specifications I.* Springer Verlag, Berlin 1985

[Genesereth, Nilsson 87]   M. R. Genesereth, N. J. Nilsson: *Logical Foundations of Artificial Intelligence* Morgan Kaufmann Publishing, Los Altos 1987

[Goguen, Burstall 85]   J. A. Goguen, R. M Burstall: *Institutions: Abstract Model Theory for Computer Science.* SRI International Report 1985

[Goguen, Meseguer 87]   J. A. Goguen, J. Meseguer: *Order-Sorted Algebra I.* SRI International Report 1987

[Herzog et al. 86]   O. Herzog et al.: *LILOG – Linguistische und logische Methoden für das maschinelle Verstehen des Deutschen.* IBM Deutschland GmbH, LILOG-Report 1a, 1986

[Kasper, Rounds 86]   R. T. Kasper, W. C. Rounds: *A Logicical Semantics for Feature Structures.* Proc. 24th ACL Meeting, 257-265, Columbia University 1986

[Luck et al. 87]   K. v. Luck, B. Nebel, C. Peltason, A. Schmiedel *The Anatomy of The BACK-System.* TU Berlin, KIT-Report No. 41, Jan. 1987

[Luck, Owsnicki-Klewe 89]   K. v. Luck, Bernd Owsnicki-Klewe: *Neuere KI-Formalismen zur Repräsentation von Wissen.* in: Christaller (Hrsg.), Künstliche Intelligenz, Springer Verlag, Berlin 1989, 157-187

[Moore 82]   R. C. Moore: *The Role of Logic in Knowledge Representation and Commonsense Reasoning* Proc. AAAI-82

[Nebel 89]   B. Nebel *Reasoning and Revision in Hybrid Reasoning Systems.* Ph. D. Thesis, University of Saarbrücken, 1989

[Nebel, Smolka 89] B. Nebel, G. Smolka: *Representation and Reasoning with Attributive Descriptions.* Proc. Workshop 'Order-Sorted Knowledge Representation and Inference Systems', Lecture Notes in Artificial Intelligence, Springer Verlag, Berlin 1989

[Oberschelp 62] A. Oberschelp: *Untersuchungen zur mehrsortigen Quantorenlogik.* Mathematische Annalen, 145:297-333, 1962

[Patel-Schneider 84] P.F. Patel-Schneider: *Small can be Beautiful in Knowledge Representation.* Proc. IEEE Workshop on Principles of Knowledge-Based Systems, 1984, 11-19

[Pletat 90] U. Pletat: *Model Theoretic Aspects of L_{LILOG}* IBM Germany, IWBS Technical Report, to appear 1990

[Pletat, Luck 89] U. Pletat, K. v. Luck: *Knowledge Representation in LILOG.* Proc. Workshop 'Order-Sorted Knowledge Representation and Inference Systems', Lecture Notes in Artificial Intelligence, Springer Verlag, Berlin 1989

[Robinson 65] J. A. Robinson: *A Macine-Oriented Logic Based on The Resolution Principle.* Journal of the ACM, Vol 12, 1965

[Schmidt-Schauss 88] M. Schmidt-Schauss: *Subsumption in KL-ONE is Undecidable.* Universität Kaiserslautern, SEKI-Report SR-88-14, 1988

[Shieber 86] S. Shieber: *An Introduction to Unification-Based Approaches to Grammar.* CSLI Lecture Notes 4, Stanford University 1986

[Smolka 88] G. Smolka: *A Feature Logic with Subsorts.* Wissenschaftliches Zentrum der IBM Deutschland, LILOG-Report 33, 1988

[Vilain 85] M. Vilain: *The Restricted Language Architecture of a Hybrid Representation System.* Proc. IJCAI-85, 1985, 547-551

[Walther 87] C. Walther: *A Many-Sorted Calculus Based on Resolution and Paramodulation.* Research Notes in Artificial Intelligence, 1987

UNIFICATION BASED MACHINE TRANSLATION

Christian Rohrer
Institut für Maschinelle Sprachverarbeitung, Universität Stuttgart
Keplerstraße 17, D-7000 Stuttgart 1

Introduction

In this paper we present an approach to MT that offers
linguistically well-motivated transfer component. We formulat
contrastive transfer rules that extend in a natural way th
independently motivated grammars of the source and target languages
Our proposal is based on Lexical Functional Grammar (LFG).
This means that we can use our LFG system, which parses sentences b
applying context-free phrase structure rules and solving equation
associated with these rules, in order to construct the target f
structure. We have only to extend our grammar and the dictionar
entries. We encode with each lexical item its translatio
equivalent. These equivalences can be quite complex. In the talk w
will illustrate our proposal with examples that have caused problem
to some other approaches. For generation we also use an LFG
Preliminary experiments using the same grammar for analysis an
synthesis have been quite successful.
LFG is a unification-based grammar formalism. Unification base
grammars are also used in other MT projects. Thus ATR (Advance
Telecommunications Research Institute International, Osaka, Japan
uses Japanese Phrase Structure Grammar and Head driven Phras
Structure Grammar for the (automatic) translation from Japanese t
English. LFG is also used at CMU (Pittsburgh) and IBM Japan for th
same language pair.
Our proposal has to be seen in connection with the move toward
lexically based grammars. Theoretical linguists have realized tha
the lexicon is the central component of grammar. Independentl
researchers in natural language processing and artificia

ntelligence have rediscovered the importance of dictionaries and
hesauri. Large scale electronic dictionaries containing
omprehensive language data, structured to be used easily in a
omputer, are being written. [0]

nification grammars

nification grammar started with Martin Kay. In parallel with the
evelopment of logic programming and logic grammars, M. Kay
eveloped a notion of structure unification which has many elements
n common with the notion of unification of first order terms. A
entral concept of unification grammar is the notion of "feature" or
attribute". Linguists have been using features for more than 50
ears. They are used in phonetics, morphology, syntax and semantics.
hus the Latin noun "murus" (wall) is characterized by the following
ttribute value pairs, in (Ia), the simplified schema for "John
icked Bill" is represented by the attribute-value pairs in (Ib).

Ia)
```
      Pred     'mur'            (Ib)   Relation  'kick'
      Gender masculine                 Argument.1 John
      Number singular                  Argument.2 Bill
      Case    nominative
```

nification is an operation which, when applied to two compatible
ttribute-value structures, yields a new attribute-value structure.
his resulting attribute-value structure contains all the
nformation of the two original attribute-value structures.
iven the dominant status of attributes, unification based grammars
re also called attribute-value based theories of grammar. The most
mportant representatives are: Lexical Functional Grammar (LFG),
eneralized Phrase Structure Grammar (GPSG), Head-driven Phrase
tructure Grammar (HPSG), Japanese Phrase Structure Grammar (JPSG),
ategorial Unification Grammar (CUG), Unification Categorial
rammar (UCG).
hy do most computational linguistics (at least in research
nvironments) work with unification grammars? In what respect do
nification grammars differ from other theories of grammar? For

unification grammars there exist general mechanisms by which the knowledge of language embodied in a particular grammar can be applied. Such mechanisms are called <u>parser-generators</u>. At the Institute of Natural Language Processing in Stuttgart we have developed a parser-generator for LFG [1]. Given an LFG for a language like German, French or Japanese, this device generates a parser for the strings specified by that grammar. Furthermore we have implemented a generator-generator which produces for an arbitrary lexical functional grammar a program which generates from a given f-structure all the grammatically well-formed surface structures [2].

All the unification grammars mentioned above are constructed from essentially the same formal devices [3]. Nevertheless there are, from a linguistic point of view, considerable differences between them. We therefore have to justify why we use LFG and, more specifically, why we think that LFG is a useful framework for machine translation. We begin with a short description of LFG. We put the main emphasis on those aspects which are particularly relevant for MT.

Structural correspondence: Mappings between constituent structure, functional structure and semantic structure

LFG assigns to every sentence two levels of syntactic representation, a constituent structure (c-structure) and a functional structure (f-structure). The constituent structure represents the surface order of the lexical items and their groupings into (larger) phrases. The lexical items are not represented as independent nodes but appear as annotations of terminal nodes of the constituent structure tree. The functional structure makes explicit the grammatical function of a sentence (subject, object, adjunct, etc). For each type of structure there exists a special notation. Constituent structures are described by standard context-free rules. Functional structures are described by Boolean combinations of function-argument equalities stated over variables which denote nodes in the c-structure.

Let us look at an example:

194

ith the context-free rule (a) we associate the equalities stated in
b).

2a) S → NP   VP

2b) $x_0 = x_2$ and $x_0(SUBJ) = x_1$

he variable x refers to S (the mother mode), x refers to NP, etc.
n standard LFG notation (a) and (b) would be collapsed into one
ule.

3) S → NP   VP

      (↑ SUBJ)=↓ ↑=↓

function ϕ maps the nodes of the constituent structure into the
odes of the functional structure [4].

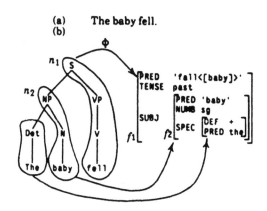

(a) The baby fell.
(b)

c-structure   f-structure

he set of equations associated with the nodes of the c-structure
onstitute a __functional-description__. The functional-structure
ssigned to a sentence is the smallest functional-structure that
atisfies the conjunction of equations in its functional
escription.

he correspondence between c-structure and f-structure was first
escribed by Bresnan and Kaplan in 1982. In the meantime several
xtensions have been proposed. Thus Kaplan [5] and Halvorsen [6]
ormulate correspondences that describe semantic schemata which are
ore abstract than f-structures. A function δ maps f-structure units
nto corresponding units of semantic structure.

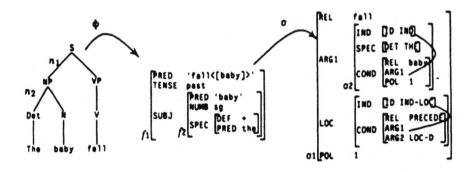

This sequence of correspondences opens up a new descriptiv
possibility. If σ and ϕ are both structural correspondences, the
so is their composition. Therefore we can formulate semanti
descriptions in terms of c-structure nodes. The expression σ (ϕ (
)) can be associated with the category x in a c-structure rule. Thi
expression denotes the semantic structure which corresponds to th
f-structure which corresponds to the c-structure node $x_\wedge$.

Correspondences between f-structures of different languages (th transfer problem)

We just have seen that LFG allows us to state correspondence
between different structures in a unified formalism. So far thes
mechanisms were used to relate different levels of structure withi
a single language. Why not use these mechanisms to relate structure
of sentences in different languages? A first step in this directio
is described in Kaplan [4]. Instead of formulating equations whic
map f-structures into semantic structures we introduce annotatior
which map f-structures of the source language into f-structures c
the target language. Since the level of semantic structure is als
at our disposal, we can view translation as correspondences on tw
levels.

SOURCE   TARGET

τ'

τ semantic structure

σ σ functional structure

$\emptyset$ $\emptyset$ constituent structure

ᴉe correspondence τ maps the f-structure of the source language
ᴉto the f-structures of the target language. τ' performs the same
ᴐeration on the semantic level.

ᴇt us consider some concrete examples. The English sentence

ᴉll gave Mary a book.

ᴀn be translated into Japanese in six different ways. Any order
ᴀong the three arguments of the verb is possible. Only the place of
ᴉe verb is fixed. It has to come at the end.

ᴅa) Bill-ga Mary-ni hon-o ageta

ᴅb) Bill-ga hon-o Mary-ni ageta

ᴅc) Mary-ni Bill-ga hon-o ageta

ᴅd) Mary-ni hon-o Bill-ga ageta

ᴅe) Hon-o Bill-ga Mary-ni ageta

ᴅf) Hon-o Mary-ni Bill-ga ageta

ᴉ LFG we relate the English verb to the corresponding Japanese verb
ᵧ the following bilingual dictionary entry.

↑ PRED) = 'geben < (SUBJ) (OBJ) OBJ2) >'

↑ τ PRED) = 'ageru <(SUBJ) (OBJ) (OBJ2) >'

↑ SUB τ) = (↑ τ SUBJ)

↑ OBJ τ) = (↑ τ OBJ)

↑ OBJ2 τ) = (↑ τ OBJ2)

more intuitive representation of the mapping   looks as follows:

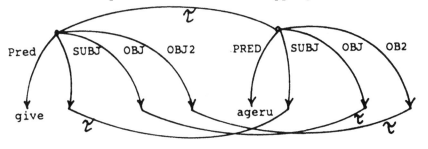

In an LFG for Japanese the 6 sentences above receive the same f
structure. In other words the f-structure of the English sentence i
mapped onto one f-structure in Japanese. From this f-structure th
six surface structures are generated by the LFG based generator. Th
fact that f-structures are more general facilitates transfer. I
transfer were formulated on constituent structures, one would nee
six transfer rules for the above example.

Of course it is not always the case that a German ver
subcategorizes the same number of syntactic functions as th
Japanese verb. The verb "verletzen" (to hurt) for instance has n
exact equivalent in Japanese. There is no transitive verb whic
means "to hurt". If one wants to translate the sentence

(5a) Marias Bemerkung verletzte Hans.

     Maria's remark hurt Hans.

one has to pick an intransitive verb. The object of "verletzen
becomes the subject of this intransitive verb. The subject o
"verletzen" is translated as an adjunct.

(5b) Hans-ga Maria-no kotoba-de kizutsui-ta.

           remark   hurt   PAST

          by Maria's remark

Similar cases occur in translating German to French. Consider th
following pair of sentences.

(6a) Hans mag Maria.

(6b) Maria plaît à Hans.

     Hans likes Maria.

Here the German subject is mapped into an à-object and the Germa
object becomes the subject in French. These relations are expresse
by the corresponding lexical entry

($\uparrow$ PRED) = 'mögen <(SUBJ)(OBJ)>'

($\uparrow\tau$ PRED) = 'plaire <(SUBJ)(à-OBJ)>'

($\uparrow$ SUBJ τ) = ($\uparrow\tau$ à-OBJ)

($\uparrow$ OBJ τ) = ($\uparrow\tau$ SUBJ)

Sometimes a superficial observer may get the impression that a
English verb has more arguments than its corresponding Japanes
verb. A closer look, however, reveals that both verbs have the sam

unctional structure and that the two language differ only on the
urface in this respect. A case in point are zero anaphora.

ero anaphora are discourse phenomena. In Japanese in the answer to
 question like

7a) Mary-wa Bill-o yobimashita ka?

               call   PAST QUEST

   "Did Mary call Bill?"

he subject and /or the object pronoun may be dropped. In other
ords, the following answers are correct [7].

7b) Hai, Mary-wa ∅ yobimashita.

    Yes, Mary (him) called.

7c) Hai, ∅ Bill-o yobimashita.

    Yes, (she) Bill called.

7d) Hai, ∅   ∅ yobimashita.

    Yes, (she) (him) called.

n this example the pronouns may be omitted because it can be
nferred from the context who called and who was called. The lexical
ntry for "to call" in Japanese contains the information that "yobu"
s a transitive verb which takes a subject and an object.

↑ PRED) = 'call <(SUBJ) (OBJ)>'

↑ℓ PRED) = 'yobu <(SUBJ) (OBJ)>'

↑ SUBJ ℓ) = (↑ℓ SUBJ)

↑ OBJ ℓ) = (↑ℓ OBJ)

he transfer rule from English (or German) to Japanese is therefore
ery simple. The omission of the pronouns must be handled by the
apanese generator (or parser, if Japanese is the source language).

exical rules

ero anaphora must not be confused with lexical rules like object
eletion.

onsider the two sentences

8a) The children ate pizza.

8b) The children ate.

n (b) the object (patient argument) is unexpressed. The unexpressed
bject does not refer to the pizza of the preceding sentence. The

sentence (b) can be paraphrased as "The children ate something"
Lexical rules relate the verbs in the following sentence pair.
(9a) John broke the vase.
(9b) The vase broke.

Transfer on semantic structures

Instead of formulating correspondences between two languages i
terms of grammatical functions like subject and object we also hav
the option of to use thematic roles. Consider the following pair o
sentences
(10a) X beschreibt die Platte mit Y.
(10b) X writes Y on the disk.
On the level of cases or thematic roles the German and th
corresponding English verb are identical.
(10c) schreiben <Agent, (Theme), (Location)>
      write   <Agent, (Theme), (Location)>
The mapping of these thematic relations on grammatical functions i
accomplished by language specific rules.
Transfer on the level of thematic roles will also be helpful fo
translating nominalizations.
(11a) John-ga Ainugo-o kenkyuu chuu.
      "during John's research in Ainu language"
On the level of thematic functions "research" and "kenkyuu" bot
subcategorize an agent and a theme
(11b) research <agent, theme>
      kenkyuun <agent, theme>
The grammatical functions in their nominalized forms differ howeve
widely between the two languages.
Transfer on semantic structures is also very useful to expres
connections between relative clauses and participial constructions
In the following example a participial construction modifies a noun.
(12a) ein vom Assistenten beantworteter Brief.
      a by the assistant answered letter.
The only possible translation into French is via a relative clause.
(12b) une lettre à laquelle l'assistant a répondu.
      a letter which the assistant answered.

he LFG system implemented in Stuttgart assigns sentence (12a) the
emantic structure represented in (12c).

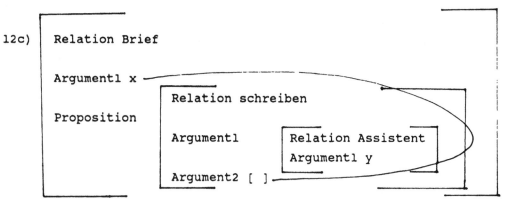

his structure subsumes the participial and the relative clause
ersion. It is abstract enough to allow a simple transfer from
erman to French (and to Japanese). The generation of surface
tructures in the target language is based on the principle of
tructure-driven derivation, described in Wedekind [2].

hat is the best level for transfer?

s you have seen our system allows transfer on two levels,
unctional and/or semantic structure. In many cases f-structure
ontains enough information for translation. Many problems which are
otoriously difficult in semantics of natural languages (like
uantifier scope) can be largely ignored in MT. In Fenstad [8] the
uthors point out correctly that the English sentence "Every
oliceman chased a man who shot a candidate." has three readings,
epending on the different scope of the quantifiers "every" and "a".
or translation into German or French these ambiguities need not to
e resolved, because we can preserve the ambiguity in the target
anguage.
n ideal MT system should translate as much as possible on the level
f f-structure. For translation between closely related languages
ike Italian or Spanish this should be fairly easy. If the f-
tructure does not contain enough information or if translation on
he level of f-strucure would involve complex transfer operations,

then one should operate on semantic structures. But even there one
could introduce several finer levels. The level of situation
schemata for instance is more abstract than f-structure but does not
yet resolve scope ambiguities. Only when it is necessary to draw
inferences (e.g. in cases of anaphora resolution) do we need a fully
explicit semantic representation like discourse representation
structures [9].

One of the challenging questions is: How does the system know when
it has to go to a higher level for transfer? We intend to
investigate this problem in the new Sonderforschungsbereich in
Stuttgart.

Disambiguation

One of the major problems in MT is disambiguation. Consider the
following example from the LILOG project.

(13) Wir können den Betrieb der Schleuse verfolgen.

   We can observe the functioning of the sluice.

The verb "verfolgen" has at least two meanings:

(14) chase, persecute

(15) observe

How can the system choose the correct word in the target language
(assuming we are translating from German to English)? Notice that
the problem of disambiguation arises already in monolingual natural
language processing. An adequate dictionary entry for "verfolgen"
would state that this verb has two readings. In the first reading
("chase") one would add the restriction that the direct object of
the verb denotes an "animate object" whereas in the second reading
the direct object denotes a "propositional entity". In knowledge-
based systems such "sortal" restrictions are associated with the
arguments of verbs, nouns, adjectives and prepositions. The sorts
are organized in a lattice which allows efficient processing. [10]
With these mechanisms one can handle ambiguities like

(16) Die Königsallee verläuft in nordsüdlicher Richtung.

   The Königsallee runs in the direction from North to South.

A bilingual dictionary gives several translation equivalents for
"verlaufen". With the help of sortal restrictions like "liquid" (das

asser verläuft, the water runs out) we can in most cases pick the correct translation equivalent.

The disambiguation of verb readings is easier than the choice of the correct reading of a preposition. In our text we find three occurrences of the preposition "nach".

17) Wenden wir uns $nach_1$ Osten, so erreichen wir $nach_2$
    Turning   East   we reach   after
    wenigen Metern den Schadowplatz.
    a few meters the Schadowplatz.

18) Die Statue wurde $nach_3$ einem Entwurf von X realisiert.
    The statue was built from a sketch by X.

In French each of the three occurrences of "nach" is translated differently.

19) $nach_1$ = vers, $nach_2$ = après, $nach_3$ = d'après

In English we also get three different surface realisations. We are not sure whether one should in general postulate for each word in the source language as many different readings as there are translation equivalents in the target language. In the case of "nach" illustrated above, however, we clearly have three different readings. Unfortunately, it is not so easy to pick out the correct reading by using sortal restrictions. The fact that "Osten" denotes a point of the compass doesn't imply that "nach Osten" always indicates a direction. Following the needle of the compass we can say

20) nach Osten kommt Süden
    after the East comes the South.

The traditional dictionaries correctly point out that "nach" is translated as "after" if it expresses a sequence ("Reihenfolge").
Therefore we can also construct a context where the word "Entwurf" does not guarantee the meaning of "$nach_3$", but also expresses a sequence ("Reihenfolge").

21) Nach dem Entwurf von Architekt X wurde der Entwurf von
    Architekt Y diskutiert.
    After the sketch by architect X we discussed the sketch by
    architect Y.

The resolution of syntactic ambiguities causes even greater problems. Let us look at an example:

22) Wir beginnen unseren Bummel über die Königsallee am Hofgarten.
    We start our stroll along the Königsallee at the Hofgarten.

This sentence illustrates the problem of pp-attachment.
Does the pp "am Hofgarten" modify the noun "Königsallee" or is "am
Hofgarten" an adjunct at the sentence level? Both readings are
possible, but the reading where the stroll along the Königsallee
begins at the Hofgarten is certainly the most plausible one.
Translating into English we can preserve the ambiguity because we
can keep the same word order. In this case MT is easier than Natural
Language Understanding. Given the sentence (22). A
question/answering-system should be able to answer the question:
(23) Wo beginnen wir unseren Bummel?
The answer we don't want is
(24) Über die Königsallee am Hofgarten.
The reading on which (24) is based can however be excluded quite
easily. The nominalization "Bummel" subcategorizes a directional
element. This is indicated by the accusative case following the
preposition "über".
A further case where the ambiguity can be preserved in English is
the complex np
(25) die Altstadt mit ihren engen Gassen und Kneipen
     the old town with its narrow lanes and pubs
Does the adjective "eng" modify only one noun ("Gassen") or does its
scope extend over the whole conjunction? If we translate into French
we have to take a decision: we can place the adjective behind the
first noun (narrow scope) or behind the second noun. In the latter
case we create a new ambiguity, because then the adjective may have
wide scope (or narrow scope with respect to the second noun).
(25a) des ruelles étroites et des bistros,
      des ruelles et bistros étroits
The problem is aggravated by the pronoun "ihr". If "eng" has narrow
scope "ihr" may have wide or narrow scope too.
For a linguist fond of ambiguities the sentence of which np (24) is
a part of presents even further possibilities of interpretation.
(26) Die Theodor-Körner-Straße führt in Düsseldorfs Altstadt mit
     ihren engen Gassen und Kneipen.
The obvious interpretation of this sentence is "the T.-K.-Street
leads to the old part of Düsseldorf." Since "Düsseldorfs Altstadt"
can be accusative or dative, one can also get the reading "In the
old part of Düsseldorf the T.-K.-Street leads (is very popular) with
its pubs..."

hen we picked the text for the LILOG project we did not try to find
 text with a maximum of ambiguities. We just picked an ordinary
ext. Even texts which are intended to be very clear, like the
ection "useful information for passengers" in the time table of the
erman Federal Railway is full of ambiguities.

27) Fahrausweise für Hin- und Rückfahrt bis 100 km
    tickets for single and return trips up to 100 kms
    gelten an dem auf dem Fahrausweis angegebenen Tag.
    are valid the day indicated on the ticket.

oes one have to add single and return trip or can each direction be
p to 100 kms long?

hat is the scope of "bis einschließlich 26 Jahre" in the next
xample?

28) Studierende und Schüler bis einschließlich 26 Jahre
    University students and highschool students up to 26

f the Deutsche Bundesbahn had an expert system covering useful
nformation for passengers it might also be used for disambiguating
ource language texts in MT. Such a proposal was made by Christian
oitet [11].

o far nobody has tried to apply this idea in a practical MT system.
 more realistic approach to the treatment of syntactic ambiguity in
T presupposes a contrastive analysis: which ambiguities can be
arried over from one language to another. Furthermore, while the
uthor is writing the source text, a parser should already point out
o him ambiguities which may be harmful even independently of
ranslation. MT can only work in the near future if we regiment the
anguage of the author and restrict the domain of application.

BIBLIOGRAPHY

[0] Research presented in this paper is financed by the BMFT grant Nos 1013212 4 and 1013210 2 as well as by IBM Germany (LILOG project). Our approach has benefited from extensive discussion with the LFG group at XEROX Palo Alto Research Center, USA. The LFG system and the generator have been implemented in Quintus Prolog on Sun workstations.

[1] Eisele, A.; Dörre, J.: A Lexical Functional Grammar System in Prolog, in Proceedings of COLING 86, Bonn 1986

[2] Wedekind, J.: Generation as Structure-Driven Derivation, in Proceedings of COLING 88, Budapest 1988

[3] For a description of the formal properties see Johnson, M.: Attribute-Value Logic and the Theory of Grammar. Stanford: CSLI, 1988

[4] Kaplan, R.M., Netter, K., Wedekind,J., Zaenen, A.: Translation by Structural Correspondence. Presented at the E-ACL Conference, Manchester, GB, April 10-12, 1989

[5] Halvorsen, P.-K.: Situation Semantics and Semantic Interpretation in Constraint-Based Grammars. In: Proceedings of the International Conference on Fifth Generation Computer Systems, Tokyo 1988

[6] Kaplan, R.: Three Seductions of Computational Psycholinguistics. In Whitelock, P., et al. (eds.): Linguistic Theory and Computer Applications, New York 1987

[7] Kameyama, M.: Zero Anaphora: The Case of Japanese, Ph. D. Dissertation, Stanford USA, 1985

[8] Fenstad, J.E., Halvorsen, P.-K., Langholm, T., van Benthem, J.: Situations, Language and Logic. Dordrecht 1987

[9] Kamp, H.: A Theory of Truth and Semantic Representation. In: Groenendijk, J. A. G., et al. (eds.): Truth, Interpretation and Information, Dordrecht 1981

[10] Smolka, G.: A Feature Logic with Subsorts, LILOG-Report 33, 1988 and Gust, H. et al.: Die Struktur des Lexikons für LILOG. LILOG-Report 29, 1988

[11] Boitet, Chr., Gerber, R., Expert Systems and other new techniques in MT, Coling 1984, Stanford, p. 468-471

Perspectives in Multiple-Valued Logic

Peter H. Schmitt
Institut für Logik, Komplexität
und Deduktionssysteme
Universität Karlsruhe

Ever since the invention of multiple-valued logic in the twenties of this century this subject has attracted a growing number of researchers. The literature in this field has been called by one author "ridiculously large" and he may well be right considering the fact that multiple-valued logic has never come close to the dominating role of classical two-valued logic. Even among non-classical logics, like modal, intuitionistic and temporal logic and more recently non-monotonic logics the multiple-valued variety seems to be regarded with marginal importance. The Helsinki Logic Machine [HLM 87], for example, provides implementations of over 60 logical calculi, but a three-valued system, let alone a multiple-valued system, is not among them.

The reason for this state of affairs is certainly not to be found in the theory of multiple-valued logic itself. An impressive body of results on axiomatizations, completeness of deduction systems, decidability, algebraic properties and many topics more has been accumulated. Also the fact that formulas in multiple-valued logic can be transformed rather straight forwardly into classical two-valued formulas does not explain the lack of general interest in this area. Though this transformation does preserve the meaning of formulas it destroys or distorts much of the relevant syntactic structure and intuition of the multiple-valued formulation.

The major obstacle to a wider recognition of the field is certainly its lack of convincing applications. The decisive question has been used by Dana Scott as the title of his delightful paper [Scott 76]: Does Many-Valued Logic Have Any Use? Scott's answer is rather sceptical. None of the proposed semantic interpretations of the additional truth values as degrees of error, as probabilities or possibilities has been very convincing. It should however be mentioned that a variation of this idea gave rise to the field of fuzzy logic, which despite of the fact that the question "where do the numbers, or rather, numerical functions come from?" is not answered satisfactorily, has found a broader dissemination. More detailed comments on fuzzy logic are beyond the scope of this paper. Even the proclamation of Jan Lukasiewicz, one of the founders of three-valued logic,

"I maintain that there are propositions which are neither true nor false but indeterminate. All sentences about future facts which are not yet decided belong to this category ..." [Lukasiewicz]

has quickly lost its impetus. Already 1975 Robert Wolf stated in his survey articles [Wolf 75] that "the philosophical discussion in recent years has become increasingly critical of the philosophical applicability of many-valued logic ...". The same may still be said today. But what about applications outside philosophy, outside of other foundational disciplines?

Since 1970 the annual International Symposium on Multiple-Valued Logic under the auspices of the IEEE Computer Society has been held with unrelenting success. The main force behind this event is the use of multiple-valued logic in electronic circuits, both on the hardware level and as a tool in the design and analysis of binary circuits. Though it should be emphasized that this focal point has never restricted the range of topics presented at the Symposia[1]. Multiple-Valued Logic circuit design is in the same position as the theory of multiple-valued logic in general: It is a minority position in the dominantly binary world. While nobody expects this situation to change drastically the advantages of multiple-valued logic hardware are more and more recognized. The most striking evidence for this trend are the commercially available numeric coprocessor Intel 8087 and the microprocessor iAPX-432 both equipped with a four-valued read-only memory, [Smith 88]. Commercial use may certainly be a powerful motivation for research but I doubt whether it was this kind of use Dana Scott had in mind.

The interest in non-classical logics including many-valued logics has been rekindled by the upsurge of the new discipline of artificial intelligence, see e.g. [Turner 85]. While philosophy, mathematics and other highly formalized sciences investigate idealized situations artificial intelligence tries to handle more common place phenomena. In theory we may assume that all facts are known, but practically we encounter situations where some facts are known to be true, others are known to be false and for the rest there is no information available. This leads us naturally to a tree-valued logic where a formula is assigned the third truth value it it is neither known to be true nor known to be false. Surprisingly enough S. C. Kleene introduced his version of three-valued logic exactly for this purpose. In [Kleene 38] he used predicates p(n), that are true, when the result of the computation of some particular Turing machine M associated with p halts on input n and outputs 0. p(n) is false, when M halts with an output ≠ 0 and p(n) is undefined when M does not halt on input n. As the title of Kleene's paper shows

---

[1] See table 1 for a classification of conference papers.

Classification of papers presented at the International Symposium on Multiple-Valued Logic

Topic	1971	1973	1975	1976	1985	1986
Switching Circuits and Logic Design	11	13	14	26	30	25
Algebra	4	4	12	7	9	9
Logic	0	0	7	2	4	4
Fuzzy Logic	2	1	5	8	10	3
Miscalaneous	1	2	12	2	3	0
Total	18	20	50	45	56	41

Table 1

the main subject of the paper was something quite different; three-valued logic only played an auxiliary role, a convenient mechanism that correctly propagated via accordingly chosen truth-tables the effects of non-terminating computations.

Basic Definitions and Notation

The syntax of the system L_3 of three-valued logic is the same as for the ordinary two-valued first order predicate calculus with the only exception, that it contains two negation symbols: ¬A for the strong negation of proposition A, ~A for weak negation. We use ∨, ∧ to denote disjunction and conjunction respectively and ∃x A, ∀x A for the existential resp. universal quantification of formula A with respect to the variable x. A structure M consists of a universe M with interpretations of the function and constant symbols as usual and a valuation function v, if necessary denoted more precisely v_M, which associates with every n-ary predicate symbol P and every n-tupel $m_1 \ldots m_n$ of elements from M a truth value $v(P(m_1 \ldots m_n))$. The possible truth values are 0 (false), 1 (true), u (undefined). The valuation function v can be extended to all formulas of L_3 without free variables using the following rules and truth tables.

Truth table for ∨			
∨	1	u	0
1	1	1	1
u	1	u	u
0	1	u	0

Truth table for ∧			
∧	1	u	0
1	1	u	0
u	u	u	0
0	0	0	0

Rules for quantifiers:

* $v(\forall x\ A) = 1$ iff for all elements m in M $v(A(m/x)) = 1$.
* $v(\forall x\ A) = 0$ iff there is an element m in M satisfying $v(A(m/x)) = 0$.
* $v(\forall x\ A) = u$ in all other cases.

* $v(\exists x\ A) = 1$ iff there is an element m in M satisfying $v(A(m/x)) = 1$.
* $v(\exists x\ A) = 0$ iff for all elements m in M $v(A(m/x)) = 0$.
* $v(\exists x\ A) = u$ in all other cases.

For the sake of this definition we have assumed that constant symbols for all elements of M are available. A(m/x) is the formula that arises from A by substituting the free occurrences of the variable x by a constant symbol for m.

Rules for ∧ and ∨

* $v(A \land B)$ = the minimum of $v(A)$ and $v(B)$.
* $v(A \lor B)$ = the maximum of $v(A)$ and $v(B)$.

Here we think of the truth values being ordered as $0 < u < 1$.

Forms of Negation

A	¬A	~A	⌐A	~A	¬¬A	⌐A
1	0	0	1	1	1	1
u	u	1	1	0	u	0
0	1	1	0	0	0	0

The content of this table will be more intuitive when rephrased in natural language:

$v(\quad A) = 1$ iff   A is true

$v(\neg A) = 1$ iff   A is false

$v(\sim A) = 1$ iff   A is not true

$v(\sim\neg A) = 1$ iff   A is not false

We abbreviate $\sim\neg A$ by ∇A.

We will use $A \supset B$ (A implies B) and $A \equiv B$ (A ist equivalent to B) as abbreviations for $\sim A \lor B$ and $(A \supset B) \land (B \supset A)$. For the convenience of the reader we give the truth tables for these new connectives:

Defined propositional operators

			B	
	A $\supset$ B	1	u	0
A	1	1	u	0
	u	1	1	1
	0	1	1	1

A formula A is called a logical consequence of a set Σ of L_3-formulas if for all L_3-structures (M, v), such that $v(B)=1$ for all B in Σ also $v(A)=1$.

A $\equiv$ B	1	u	0
1	1	u	0
u	u	1	1
0	0	1	1

A formula A is called an L_3-tautology if for all L_3-structures (M, v) we have $v(1)=1$.

Tableau calculus for L_3

When E. W. Beth presented in 1958 his tableau calculus to the International Congress of Mathematicians he remarked:

"With a view to many-valued logic this possibility [the possibility to find a correct and complete extension of his tableau calculus for many-valued logic] is so obvious as to make any further discussion superfluous."

Despite Beth's remark or maybe because of it it took almost 30 years till W. A. Carnielli finally did the job.

In this chapter we present a sound and complete tableau calculus for L_3, which is derived from [Carnielli 87]. Carnielli's paper is tremendously general. It contains Tableau calculi for propositional and first-order many-valued logics with arbitrary finite number of truth values, arbitrary set of designated truth valued, arbitrary finitary propositional connectives and arbitrary distribution quantifiers. Our system may be obtained as a special case of Carnielli's method except for the fact, that we allow function symbols, while he does not and our rules for quantifiers are much simpler than his. To illustrate his method Carnielli presents in his paper the quantifier rules for L_3 in detail, see page 489. Our simplification is mainly due to the use of the unary propositional connective ∇, which already played a crucial role in the resolution calculus for L_3 presented in [Schmitt 86].

As in [Carnielli 87] and [Smullyan 68] we use signed formulas, i. e. strings of the form TA, UA or FA where A is an arbitrary L_3-formula. For an L_3-structure (M, v) and a closed L_3-formula A we define

TA	is true in (M, v)	iff	$v(A)=1$
UA	is true in (M, v)	iff	$v(A)=u$
FA	is true in (M, v)	iff	$v(A)=0$.

We will use S as a metavariable ranging over signs; thus SA will refer to any of the signed formulas TA, UA or FA.

A tableau is a tree, as usual drawn upside down, whose nodes are labeled with signed formulas. The basic method used in the proof procedure to be described is the construction of closed tableaux. This is started by putting down one signed formula as the label of the root node and proceeds by stepwise extensions. To perform an extension step one branch b of the already constructed tableau is selected, one signed formula SA on this branch b is selected and according to the rule associated with SA b is extended. The extension of b may take one of the following two forms:

1.   Two or more signed formulas $C_1,...,C_r$ are attached at the end of b yielding one new branch b'.

2.   b splits into two or more branches $b_1,...,b_k$, where b_i is obtained by attaching signed formulas $C_{i1},...,C_{ir(i)}$ at the end of b.

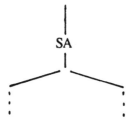

In case 1 the extension rule is written

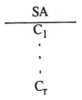

$$\frac{SA}{C_1}$$
$$\vdots$$
$$C_r$$

In case 2 the extension rule has the form

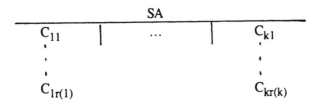

The list of extension rules is given below.

A branch b is closed if

* there are signed formulas S_1A_1 and S_2A_2 on b such that $S_1 \neq S_2$ and A_1 and A_2 are unifyable

* there is a signed formula SA on b, for which no extension rule exists.

The second alternative in this definition has no correspondence in the two valued case where extension rules for all signed formulas exist. In L_3 there is e. g. no extension rule for the formula U~A. A tableau is closed if all its branches are closed. From what has been said so far the construction of tableau can be seen to be nondeterministic. The extension rules are to be understood as possible steps, no obligation to apply an extension rule exists except for the overall effort to arrive at a closed tableau.

The main results are:

Theorem 1:   Let A be a closed L_3-formula.
A is an L_3-tautology
            iff
* there is a closed tableau with root FA
and
* there is a closed tableau with root UA.

If Σ is a set of L_3-formulas a tableau **over** Σ is constructed as above with the additional rule that we may extend at any time a branch b by attaching TB at the end of it for some formula B in Σ.

Theorem 2:   Let Σ be a set of closed L_3-formulas and A a single closed L_3-formula.

A is a logical consequence of Σ

   iff

* there is a closed tableau over Σ with root FA

and

* there is a closed tableau over Σ with root UA.

The proofs of the theorems are easily obtained along the lines of the corresponding proofs in [Smullyan 68] and [Carnielli 87] and will be published elsewhere.

The extension rules

for $\wedge$:

$$TA \wedge B$$
$$\overline{}$$
$$TA$$
$$TB$$

$$FA \wedge B$$
$$\overline{}$$
$$FA \mid FB$$

$$UA \wedge B$$
$$\overline{}$$

UA	UA	TA
TB	UB	UB

for $\vee$:

$$TA \vee B$$
$$\overline{}$$
$$TA \mid TB$$

$$FA \vee B$$
$$\overline{}$$
$$FA$$
$$FB$$

$$UA \vee B$$

UA		UA		FA
UB		FB		UB

for negations:

$T \neg A$	$F \neg A$	$U \neg A$
FA	TA	UA

$T \sim A$	$F \sim A$
FA | UA	TA

$T \nabla A$	$F \nabla A$
TA | UA	FA

for $\supset$:

$TA \supset B$	$FA \supset B$
FA | UA | TB	TA
	FB

$$UA \supset B$$

TA
UB

$T\forall x\, A(x)$	$F\forall x\, A(x)$
TA(x)	$FA(f(y_1,...,y_n))$

$$U\forall x\ A(x)$$

$$T\nabla A(x)$$
$$UA(f(y_1,...,y_n))$$

here f is a new function symbol not occurring in the tableau before and $y_1,...,y_n$ are all free variables of $\forall x\ A(x)$.

$$T\exists x\ A(x) \qquad\qquad F\exists x\ A(x)$$

$$TA(f(y_1,...,y_n)) \qquad\qquad FA(x)$$

$$U\exists x\ A(x)$$

$$T{\sim}A(x)$$
$$UA(f(y_1,...,y_n))$$

here f is a new function symbol and $y_1,...,y_n$ are all free variables of $\exists x\ A(x)$.

Examples

In classical two-valued logic $({\sim}B{\supset}{\sim}A) \supset (A{\supset}B)$ is a tautology. The following two closed tableaux show that it remains a tautology in three-valued logic.

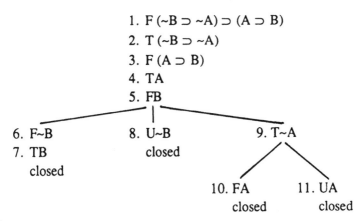

1. F $({\sim}B \supset {\sim}A) \supset (A \supset B)$
2. T $({\sim}B \supset {\sim}A)$
3. F $(A \supset B)$
4. TA
5. FB

6. F$\sim$B   8. U$\sim$B   9. T$\sim$A
7. TB   closed
   closed

10. FA   11. UA
    closed   closed

Tableau 1

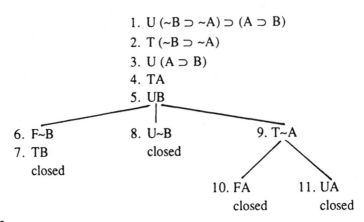

Tableau 2

The formula obtained by replacing weak negation by strong negation $(\neg B \supset \neg A) \supset (A \supset B)$ on the other hand is not a three-valued tautology. The tableau for $F(\neg B \supset \neg A) \supset (A \supset B)$ still closes but for $U(\neg B \supset \neg A) \supset (A \supset B)$ the following open complete tableau is obtained:

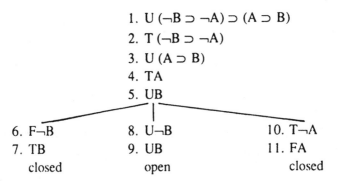

Tableau 3

As a more difficult example let us prove that $B = \exists x \, \forall y \, A(x,y) \supset \forall y \, \exists x \, A(x,y)$ is also a three-valued tautology. FB yields a simple non-branching closed tableau, so let us look at UB.

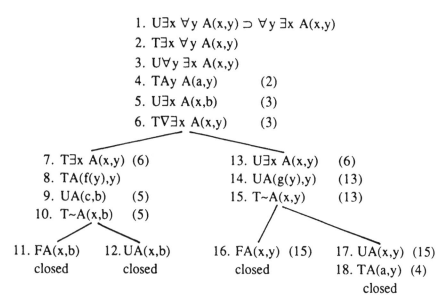

1. $U \exists x \, \forall y \, A(x,y) \supset \forall y \, \exists x \, A(x,y)$
2. $T \exists x \, \forall y \, A(x,y)$
3. $U \forall y \, \exists x \, A(x,y)$
4. $T \forall y \, A(a,y)$ (2)
5. $U \exists x \, A(x,b)$ (3)
6. $T \forall \exists x \, A(x,y)$ (3)

7. $T \exists x \, A(x,y)$ (6)   13. $U \exists x \, A(x,y)$ (6)
8. $TA(f(y),y)$ 14. $UA(g(y),y)$ (13)
9. $UA(c,b)$ (5)   15. $T \sim A(x,y)$ (13)
10. $T \sim A(x,b)$ (5)

11. $FA(x,b)$ 12. $UA(x,b)$ 16. $FA(x,y)$ (15)   17. $UA(x,y)$ (15)
closed   closed   closed   18. $TA(a,y)$ (4)
                                                           closed

Tableau 4

Conclusion

At present no convincing use of finitely many-valued logic on a larger scale can be observed. But there are hopeful potentials in semantics of natural language, in dialogue systems, in knowledge representation, in deductive database theory, that are worth trying. As a first step a flexible inference engine for three-valued logic should be available. We presented the first step towards ist implementation: the theoretical basis. The next step consists in the transformation of the theoretical calculus into an informal specification of a program realizing the inference engine. Besides the usual changes:

the use of unification in combination with the γ-rule,

the observation that all but the γ-rule need only be applied once for each formula on one branch,

there are new issues in the three-valued case:

Is it really necessary to construct two tableaux?

Is it, maybe, possible to combine them into one?

First hand calculated experiments also indicate, that the conviction backward chaining in classical logic is easier to be controlled than forward chaining may not be true for many-valued logic.

References

[Beth 58]   E. W. Beth
                 Completeness Results for Formal Systems
                 Proc. Int. Congress of Mathematics 1958

[Carnielli 87]   W. A. Carnielli
                 Systematization of Finite Many-Valued Logics Through the
                 Method of Tableau
                 J. of Symbolic Logic, 52, 1987, pp. 473 - 493

[HLM 87]   I. Niemelá, H. Tuominen
                 Helsinki Logic Machine: A System for Logical Expertise
                 Helsinki University of Technology, Digital Systems
                 Laboratory, Techn. Report, Series B, No. 1, 1987

[Kleene 38]   S. C. Kleene
                 On a notation for ordinal numbers
                 J. Symbolic Logic 3, 1938, pp. 150 - 155

[Lukasiewicz]   J. Lukasiewicz
                 On Determinism
                 in: Polish Logic 1920 - 1939
                       St. McCall (ed.), Clarendon Press, 1967, pp. 19 - 39

[Schmitt 86]   P. H. Schmitt
                 Computational aspects of three-valued logic
                 in: Siekmann, J. (ed.): Proceedings of the 8th Int. Conf. on
                       Automated Deduction, Lecture Notes in Computer Science,
                       Vol. 230, Springer 1986, pp. 190 - 198

[Scott 76]   D. Scott
                 Does Many-Valued Logic Have Any Use?
                 Philosophy of Logic. Proceeding of the 3rd Bristol
                 Conference on Critical Philosophy (S. Koerner ed.),
                 Blackwell, Oxford, 1976, pp. 64 - 95

[Smith 88]   K. C. Smith
                 Multiple-Valued Logic: A Tutorial and Appreciation
                 IEEE Computer Society, April 1988, pp. 17 - 27

[Smullyan 68]   R. M. Smullyan
                  First order logic
                  Springer 1968

[Scott]   D. Scott
                  Does Many-Valued Logic Have Any Use?

[Turner 85]   R. Turner
                  Logics for Artificial Intelligence
                  Ellis Horwood Ltd., 1985

[Wolf 75]   R. G. Wolf
                  A Critical Survey of Many-Valued Logics 1966 - 1874
                  Proc. Int. Symp. on Multiple-Valued Logic,
                  IEEE Computer Society, 1975, pp. 468 - 474

Properties and Actions

Wolfgang Schönfeld

IBM Germany Scientific Center - IWBS

Wilckensstr. 1a, D-6900 Heidelberg 1

SCHFELD at DHDIBM1 (EARN)

Abstract

One of the current issues of knowledge processing is the problem of integrating knowledge representation with classical programming languages. This may have two reasons. First, large software products already exist, and 'newcomers' have to take care of how they fit into the landscape. Second, some problems seem to inherently call for procedural programming, either because of execution speed or since their algorithmic treatment is known from other contexts. It is e.g. not appropriate to program matrix multiplication in Prolog.

In this paper, a general solution is presented for the problem of integrating full first-order predicate logic in a procedural language like PASCAL. We suspect that our approach can be used to develop a formal semantics for fragments of natural language describing properties and actions.

Introduction

Whereas in classical data processing we have programs and (explicitly specified) data, the new knowledge based techniques force us to rethink this classification of objects stored in a computer. A knowledge base may, on the one side, be viewed as implicitly specified data and, on the other side, as a non-deterministic program. It is one of the main paradigms of logic programming to view facts and rules as a whole.

This fruitful approach led to the misunderstanding that the knowledge-based techniques might replace classical programming at all. At least practicians soon realized that this cannot be true. Now they want to know the advantages of these new techniques. We have the impression that this question has not yet found a satisfactory theoretical answer and is not even fully recognized.

We are not able to present an answer here. Instead, we will develop a language concept which integrates knowledge representation in classical programming, with the tacit hope that, if the practician can choose freely, he will eventually find out which way is better and why.

Clearly, this idea is not new. Most AI languages have some procedural interface. The standard Horn clause language Prolog and its descendants (see [Muller85] for an example) have *built-in predicates*. Production rule systems like OPS5 incorporate *procedure calls* as rule consequences. In both cases, some procedures (pre-defined or user-defined) are executed which exchange data with the inference engine.

The approach presented here ([Schf89] being a predecessor paper) is new in that it copes with full first-order logic. Shortly speaking, we describe how inference engines for full first-order logic could interact with the other system components. To this end, we view results of procedure execution as (explicitly specified) data extending the pure descriptive knowledge base. Then we develop a common operational semantics concentrating on the questions

How is the inference engine called by the outer system?
How does the inference engine call the outer system?

The first question describes the standard situation of an inference engine as an application program. The second question faces the problems of procedural attachments.

A clear separation on the side of syntax will help us in defining the semantics in a satisfactorily formal way: We start with a pure logical and with a pure procedural interpretation ("either think or act") and then develop the semantics at the interface. We allow control to flow from procedure to logic, from there back to procedure, again to logic and so on. This means that we have logical $meta^n$-languages for any $n \in \mathbb{N}$. To avoid the well-known problems of mixing object and metalanguage known from [Bowen82], we require all levels to be different from each other.

Our solution may help to solve the following application problems:

- consistent updates of the knowledge base
- metaknowledge
- explanations of object and metareasoning

Since we follow both the logic programming and the production system approach as closely as possible, we get as a byproduct an amalgamation of the two.

Some basic definitions of logic

Let us shortly review some basic definitions of *mathematical logic* following [Barwise77].

Language of logic and its interpretations

A *language of logic* L is the following. Given some symbols $c_1, c_2, ...$ for constant elements, $f_1, f_2, ...$ for functions, and a set $V = \{v_1, v_2, ...\}$ of variables, the *terms* of L are constructed in the usual way. An *atomic formula* is an expression of the form $r(t_1, ..., t_n)$ where the t_i are terms and r is one of the relation symbols $r_1, r_2, ...$ and has arity n. A *literal* is an atomic formula or its negation. An atomic formula and its negation are said to be a *complementary pair* of literals. Arbitrary *well-formed formulas* of L are composed of atomic formulas using the standard logical operators $\wedge, \vee, \neg, \forall, \exists, ...$ as well as some auxiliary symbols like parentheses, commas etc.

A *structure for* L is a pair $\mathbf{A} = (A, F)$ where A is a non-empty set and F is a function which assigns

> to any constant symbol c an element $c^A \in A$,
> to any n-ary function symbol f a function $f^A : A^n \to A$,
> to any n-ary relation symbol r a relation $r^A \subseteq A^n$.

and is usually written as $\mathbf{A} = (A, r^A, ..., f^A, ..., c^A, ...)$.

An *assignment in* $\mathbf{A}$ is a function $h : V \mapsto A$. For any term t of L, we denote by $t^A[h]$ the value of t under h. If t is constant, i.e. contains no variables, we abbreviate by t^A. We write $\mathbf{A} \models \alpha[h]$ if the assignment h *satisfies* formula α in $\mathbf{A}$. The notation $\mathbf{A} \models \alpha$ means that $\mathbf{A} \models \alpha[h]$ for all assignments h in $\mathbf{A}$. We write $\mathbf{A} \models \Sigma$ for sets Σ of formulas if $\mathbf{A} \models \alpha$ for all $\alpha \in \Sigma$.

We use the symbol $\models$ also to denote consequence, i.e. $\Sigma \models \alpha$ holds if for all L-structures $\mathbf{A}$ with $\mathbf{A} \models \Sigma$ we have $\mathbf{A} \models \alpha$.

Substitutions

Let c be a constant symbol which is not yet in language L, and denote by (L, c) the extension of L by c. Then $(\mathbf{A}, a)$ denotes the (L, c)-structure which coincides with $\mathbf{A}$ on L and interpretes the additional constant c by the element $a \in A$ (i.e. $c^{(A,a)} = a$).

Let α be a formula with a single variable x. Denote by $\alpha\{x/t\}$ the result of substituting x by term t in α. For any assignment h in $\mathbf{A}$ with $h(x) = a$, we have

$$\mathbf{A} \models \alpha[h] \quad \Leftrightarrow \quad (\mathbf{A}, a) \models \alpha\{x/c\}$$

This *transformation theorem* means that we may consider substitution as a syntactic counterpart of assignment (if language L has enough constants).

Deduction

The syntactic counterpart of the semantically defined relation of consequence is that of *formal provability*. For some given calculus C, we write $\Sigma \vdash_c \alpha$ if we can prove α from Σ within C. The calculus C is *correct* and *complete* if $\Sigma \models \alpha \Leftrightarrow \Sigma \vdash_c \alpha$. This means that either $\Sigma \vdash_c \alpha$ or there is an $A \models \Sigma \cup \{\neg\alpha\}$. For our considerations, we prefer to work with a calculus which is *constructive* in both cases: Not only does it effectively generate a proof if there is one, it also generates a counterexample (possibly infinite) in the opposite case. The reason is that we want to be able to decide questions if this is at all possible.

Viewed from the outer system, there are two essentially different usages of Σ : Evaluate it to produce a result of a **query**

For a given proposition α, does $\Sigma \models \alpha$ hold ?

or of a **notification**

For a given proposition η, is $\Sigma \cup \{\eta\}$ inconsistent ?

The inference engine reduces queries to notifications via

$$\Sigma \models \alpha \quad \Leftrightarrow \quad \Sigma \cup \{\neg\alpha\} \text{ inconsistent}$$

Note that the user has different expectations: Upon a query (during a consultation), he expects that the inference engine answers "inconsistent" whereas he would be worried by that answer in the notification case. We have to take that into account when we think about calling attached procedures.

Because of this necessity to cope with failure of proof, we cannot use resolution calculus (a 'cut-only' calculus). Instead, we have to base our considerations on cut-free sequent calculus, as initiated by the development of the connection method of [Bibel83]. We use the *structure sharing variant* of sequent calculus, the so-called tableau calculus as developed in [Beth59], following the lines of [Schf85].

For the sake of brevity, we assume that any formula α is a clause, i.e. of the form $l_1 \vee \ldots \vee l_n$ where the l_i are literals. Let a knowledge base Σ be given. We consider bipartite trees T with nodes labeled either by $\alpha\sigma$ where $\alpha \in \Sigma$ and σ some substitution (*formula node*)

or by l where l is a literal in some $\alpha \in \Sigma$ (*literal node*)

The root is a formula node.

Let U be a subtree of T. By σ_U we denote the join of the σ which occur in U (if it exists). $p_1, \neg p_2$ are *complementary* σ_U if $p_1\sigma_U \equiv p_2\sigma_U$.

The *alternating tableau* of Σ, denoted by $T_A(\Sigma)$, is constructed according to the following rules:

1. Start: Choose some $\alpha \in \Sigma$ as root and its literals $l_1, ... , l_n$ as the successors.
2. Now let, in an intermediate state, T be the tree constructed so far.
3. If all branches B are contradictory, i.e. contain a pair of literals which is complementary modulo σ_B , then stop.
4. Otherwise, choose a non-contradictory branch B with leaf l.
5. For all $\alpha \in \Sigma$ and all σ such that σ is a most general unifier of the complement of l and some literal in α:
6. Append to l the node $\alpha\sigma$ with the literals of α as its immediate successors.

We assume that no two clauses built into the tableau have any universal variable in common.

Example: Let $\Sigma = \{\alpha_0, ... , \alpha_4\}$ with $\alpha_0 = p$, $\alpha_1 = \neg r \rightarrow s$, $\alpha_2 = p \wedge r \rightarrow q$, $\alpha_3 = r$, $\alpha_4 = q \rightarrow \neg p$.
Its alternating tableau is (contradictory branches marked by ■):

A *proof* is a subtree T of $T_A(\Sigma)$ such that

1. T contains the root.
2. σ_T exists.
3. Each branch of T is contradictory modulo σ_T.
4. With each formula node in T, each successor is in T.
5. With each literal node in T, exactly one successor is in T.

Example: The following subtree of the above is a proof:

$$
\frac{\begin{array}{c}\alpha_0 \\ \hline p \\ \hline \alpha_2\end{array}}{
\begin{array}{ccc}
\neg p & \neg r & q \\
\blacksquare & \dfrac{\alpha_3}{r} & \dfrac{\alpha_4}{\neg q \ \neg p} \\
 & \blacksquare & \blacksquare \quad \blacksquare
\end{array}}
$$

A *disproof* is a subtree T' of $T \in T_A(\Sigma)$ such that:

1. T' contains the root.
2. For any branch B in T', σ_B exists.
3. There are no two literals k, l in T' such $k\sigma_B$ is complementary to $l\sigma_C$ where B (resp. C) is a branch through k (resp. l).
4. With each literal node in T', each successor is in T'.
5. With each formula node in T', exactly one successor is in T'.

Note that this definition is dual to that of a proof.

Example: For $\Gamma = \Sigma - \{\alpha_3\}$, we get the (one-element) set of alternating tableaux

$$
\frac{\begin{array}{c}\alpha_0 \\ \hline p\end{array}}{
\begin{array}{cc}
\dfrac{\alpha_2}{\neg r} & \dfrac{\alpha_4}{\neg q} \\
\dfrac{\alpha_1}{s} & \dfrac{\alpha_2}{\neg r} \\
 & \dfrac{\alpha_1}{s}
\end{array}}
$$

It can be shown that $T_A(\Sigma)$ contains either a proof or a disproof. In the first case, Σ is unsatisfiable. In the second case, we can construct a $A \models \Sigma$ from the disproof.

Example: For Γ as above, we choose an A which assigns truth value *true* to p,s and truth value *false* to q,r. It is easily seen that $A \models \Gamma$.

For the rest of this paper, we assume that there is an inference engine which searches for proof/disproof according to the principle elaborated above. Note that $T_A(\Sigma)$ describes this search space. We do not specify how it is processed (breadth-first etc.).

Informal Semantics of the interface

Imagine you would talk to the computer about properties (of states) and about actions (changing states) if it were able to 'understand' natural language. Usually, you use declarative sentences to communicate to the addressee that a certain property holds, e.g. "Today is Friday, 22th of June" or that an action is ongoing, e.g. "If today is the 19th, the monthly report is written". Imperative sentences are used to force the addressee to execute an action "To write a monthly report, call the editor and type the report". Note that no sentence is both declarative and imperative. Despite this clear syntactic separation, there is (necessarily) a close semantic interconnection by the way of reference, as the above example shows: We know that the action described by "the monthly report is written" in the declarative sentence is the same as that of "To write a monthly report" in the command.

To simplify, we associate the semantics of declarative sentences with the evaluation of knowledge bases, and of imperative sentences with classical programs. The reference mechanism of the natural language will then guide the development of an interface between the two.

Formal semantics of the interface

We assume that the outer system is described in some programming language P the semantics of which is given by some set-theoretic structure in the standard denotational style (see [Stoy77]). This structure has to incorporate higher type objects (not only first-order) because of the presence of functions, functionals etc. A complete solution to the integration problem would have to take this into account. We restrict this structure to a first-order structure A , the *standard model*.

The knowledge base is a set of axioms which defines a class of models one of which should be the standard model. The knowledge base is evaluated deductively, producing an interpretation in the *term model* used in the completeness proofs (of first-order predicate logic). Interleaved with that execution, certain procedures are executed by the outer system resulting in the corresponding input-output relations of the standard model. Assuming that the logical language contains terms to denote any element of the standard model (i.e. assuming that we have its *diagram*), we can describe the input-output relation in a purely syntactical way, applying the transformation theorem. These syntactic descriptions of the input-output relation are online updates of the knowledge base so that the ongoing deduction may make use of them. Let us describe this in more detail.

Programming Language ↦ Logic

Let Σ be a knowledge base of formulas in first-order predicate logic. Evaluation of Σ takes place if a query or a notification occurs. The outer system calls the inference engine by

prove α **from** Σ **result** *id*

or by

verify α **in** Σ **result** *id*

with the appropriately defined input data types for knowledge base and formula. The possible outputs of the inference engine vary from a simple "yes" (or "no") over found bindings of universal variables to complete proofs (or disproofs). Hence, we leave open the type of the identifier *id* which is used to return the result of the call.

In addition, the inference engine must have functions which support the development of Σ :

- syntax analysis of logical objects (terms, formulas, sets of formulas, ...)
- representation and manipulation of logical objects
- representation and explanation of proofs and disproofs
- ...

The flow of control is:

1. Call a deductive procedure (query, notification)
2. Return a boolean value, proof, counterexample

Note that, by carrying out this implementation, we have extended the programming language to a metalanguage for logic.

Logic ↦ Programming Language

Assume that for each n-argument procedure p defined in P there is a *procedure predicate* $p(in{:}x_1, \dots , x_n, out{:}y_1, \dots , y_n)$ in L describing the input-output relation of p. I.e. for any assignment h in the standard model, we require that

$$A \models p(in{:}x_1, \dots , x_n, out{:}y_1, \dots , y_n)[h]$$

iff procedure p is executed with initial actual parameters $x_1{}^A[h], \dots , x_n{}^A[h]$ and final actual parameters $y_1{}^A[h], \dots , y_n{}^A[h]$. Note that this is undefined in case of an infinite loop in p.

Remark: In practice, we cannot assume that reference between L and P simply works by the name p. Instead, we have to pass the specifications of P-constructs as parameters to the inference engine, e.g.

prove α **from** Σ **using** $p_1, \dots, p_n$ **result** id

or by

verify α **in** Σ **using** $p_1, \dots, p_n$ **result** id

Reference is then be achieved by the usual parameter passing mechanism.

We assume that for each element $a \in A$ (which is an admissible attribute value in a procedure predicate), there is a constant term t in L such that $t^\wedge = a$.

The flow of control is as follows:

Suppose that the inference engine encounters some procedure predicate $p(in{:}x_1, \dots, x_n, out{:}y_1, \dots, y_n)$ with some actual substitution σ. If it decides to call the attached procedure (a prerequisite is that σ be defined for the x_v and undefined for the y_v), then:

1. For $v = 1, \dots, n$ the values of $(x_v \sigma)^\wedge$ are computed and passed as parameters to p.
2. p is executed.
3. For $v = 1, \dots, n$ a term t_v is generated such that $t_v^\wedge$ is the vth final actual parameter of p.
4. σ is extended as to assign t_v to y_v .
5. Deduction is continued with $p(in{:}x_1, \dots, x_n, out{:}y_1, \dots, y_n)\sigma$ added to the proof.

In case the inference engine decides not to call the procedure, deduction is continued with $\neg p(in{:}x_1, \dots, x_n, out{:}y_1, \dots, y_n)\sigma$ added to the proof (σ unchanged).

Note that we let open the question whether the procedure predicate or its negation is added to the knowledge base. I.e. we do not specify whether the inference engine *remembers* what it has done with the attached procedure.

We can single out the following principles for a strategy to execute attached procedures:

- It must happen in time since otherwise the knowledge base would remain unnecessarily incomplete.
- It must happen economically since it might entail unnecessary efforts and side effects.

They are satisfied if the inference engine

1. upon a query tries to generate a proof:

attempt at closing branches by contradictions

2. upon a notification tries to generate a disproof:
   avoid closing branches by contradictions

Example: Let α be the formula $q(x) \rightarrow succ(in{:}x, out{:}y)$ where q, r are usual relations and $succ$ is a procedure predicate. Suppose that the inference engine was notified that $q(4)$ holds. Then it evaluates $\alpha\sigma$ with substitution $\sigma = \{x/4\}$ and calls procedure $succ$ with input value $4^{\wedge}$. Since $5^{\wedge}$ is the output value, σ is extended as to assign the term 5 to the variable y. Predicate $succ(in{:}x, out{:}y)$ is added to the proof.

$$\frac{\neg q(x) \quad \dfrac{\dfrac{q(4)}{\alpha\{x/4\}}}{succ(x,y)}}{\blacksquare \qquad succ(x,y)\{y/5\}}$$

Remember that non-execution of $succ$ would result in an addition of $\neg succ(x,y)$ and, hence, in an unexpected inconsistency.

Suppose now that the inference engine is asked whether $\neg q(4)$ can be derived from the knowledge base. It starts search with assumption $q(4)$, unifies x with 4 and applies α. Then, procedure $succ$ is **not** called resulting in the addition of $\neg succ(in{:}x, out{:}y)$ to the current proof.

$$\frac{\neg q(x) \quad \dfrac{\dfrac{\dfrac{q(4)}{\alpha\{x/4\}}}{succ(x,y)}}{\neg succ(x,y)}}{\blacksquare \qquad\qquad \blacksquare}$$

Note that execution of $succ$ would result in an addition of $succ(x,y)$ and in the non-ability to answer the question.

Reasoning with postconditions

Suppose that Σ contains a formula

$$inverse(y,z) \leftarrow div(in{:}1, y, out{:}z)$$

If we asked the inference engine to compute the inverse of 6 by issuing the call

prove $inverse(0,z)$ **from** Σ **result** id

the inference engine would prove that by trying to execute the procedure attached to predicate $div(in{:}1, 0, out{:}z)$. Clearly, this would have unwanted side effects.

To avoid this, the knowledge base must contain the formula

$$y = 0 \rightarrow \neg \, \mathrm{div}(x,y,z)$$

meaning that the division procedure should not be executed in case $y = 0$. The inference engine, before executing div , can issue the intermediate call to itself

notify $\mathrm{div}(1,0,z)$ **from** Σ **result** *id*

and derive the contradiction $0 \neq 0$. On this basis, it can decide not to execute div since it knows in advance that execution would result in an inconsistency.

Note that we cannot expect that we avoid all such inconsistencies since this would require that we had derived **complete** knowledge about the procedure by carrying out a formal program correctness proof. Nevertheless, we claim that even incomplete procedure knowledge will help to circumvent unwanted inconsistent situations: Even if we don't know all consequences of procedure execution in advance we should at least apply what is known to avoid errors, infinite loops etc.

Possible extensions

Note that we assumed a logical language L without sorts. This means that the above version of our approach only works for an untyped programming language. To incorporate sorts seems to be not too difficult but has still to be worked out.

We believe that the interface problem is slightly more complicated if L has built-in equality: It is no longer obvious how an output value (of the standard model) should be denoted by an L-term.

An extension of the interface would be the admission of attached functions: As soon as the inference engine encounters a term t (e.g. during unification), an attached function is called to transform t to a certain normal representation. E.g. the term $3 + 4$ would be replaced by the term 7.

Let us finally remark that the language concept proposed in this paper has not yet been implemented in full detail. As experience with programming languages shows, must language designs have to be revised when they undergo the process of implementation. Nevertheless, the experiments we have carried out so far convinced us that the basics of the concept are reasonable.

References

[Barwise77]   J. Barwise (Ed.), Handbook of Mathematical Logic, North-Holland, Amsterdam 1977

[Beth59]   E.W. Beth, The foundations of mathematics, North-Holland Pub. Co., Amsterdam 1959

[Bibel83]   W. Bibel, Matings in Matrices, Communications of the ACM 26(1983), 844-852

[Bowen82]   K.A. Bowen, R.A. Kowalski, Amalgamation of language and metalanguage in logic programming, in: D.D. Clark, S.-A. Tärnlund (Eds.), Logic Programming, Academic Press (1982), 153-172

[Muller85]   C. Muller, Modula-Prolog User Manual, KLR 85-107 C, Brown, Boveri & Company, Baden/Switzerland (1985)

[Schf85]   W. Schönfeld, PROLOG extensions based on tableau calculus, Proc. 9th Int. Conf. Artificial Intelligence, Aug. 1985, Los Angeles, Ca., Vol. 2, pp. 730-732.

[Schf89]   W. Schönfeld, W. Stephan, Integration of Descriptive and Procedural Language Constructs, to appear in the Proc. Computer Science Logic '88 (Duisburg), Springer Lecture Notes in Computer Science.

[Stoy77]   J. E. Stoy, Denotational Semantics: The Scott-Strachey Approach to Programming Language Theory, MIT Press, Cambridge, Mass., 1977.

Rationale and Methods for Abductive Reasoning in Natural-Language Interpretation*

Mark E. Stickel

Artificial Intelligence Center
SRI International
Menlo Park, California 94025

Abstract

By determining those added assumptions sufficient to make the logical form of a natural-language sentence provable, abductive inference can be used in the interpretation of sentences to determine the information to be added to the listener's knowledge, i.e., what the listener should learn from the sentence. Some new forms of abduction are more appropriate to the task of interpreting natural language than those used in the traditional diagnostic and design synthesis applications of abduction. In one new form, least specific abduction, only literals in the logical form of the sentence can be assumed. The assignment of numeric costs to axioms and assumable literals permits specification of preferences on different abductive explanations. Least specific abduction is sometimes too restrictive. Better explanations can sometimes be found if literals obtained by backward chaining can also be assumed. Assumption costs for such literals are determined by the assumption costs of literals in the logical form and functions attached to the antecedents of the implications. There is a new Prolog-like inference system that computes minimum-cost explanations for these abductive reasoning methods.

1   Introduction

We introduce a Prolog-like inference system for computing minimum-cost abductive explanations. This work is being applied to the task of natural-language interpretation, but

*This research is supported by the Defense Advanced Research Projects Agency, under Contract N00014-85-C-0013 with the Office of Naval Research, and by the National Science Foundation, under Grant CCR-8611116. The views and conclusions contained herein are those of the author and should not be interpreted as necessarily representing the official policies, either expressed or implied, of the Defense Advanced Research Projects Agency, the National Science Foundation, or the United States government. Approved for public release. Distribution unlimited.

other applications abound. Abductive inference is inference to the best explanation. The process of interpreting sentences in discourse can be viewed as the process of generating the best explanation as to why a sentence is true, given what is already known [8]; this includes determining what information must be added to the listener's knowledge (what assumptions must be made) for the listener to know the sentence to be true.[1]

To appreciate the value of an abductive inference system over and above that of a merely deductive inference system, consider a Prolog specification of graduation requirements: e.g., to graduate with a computer science degree, one must fulfill the computer science, mathematics, and engineering requirements; the computer science requirements can be satisfied by taking certain courses, etc. As an example of a deductive-database application [11], the graduation requirements generate:

```
csReq <- basicCS, mathReq, advancedCS, engReq, natSciReq.
engReq <- digSys.
natSciReq <- physicsI, physicsII.
natSciReq <- chemI, chemII.
natSciReq <- bioI, bioII.
            :
```

After the addition of facts about courses a student has taken, such a database can be queried to ascertain whether the student meets the requirements for graduation. Evaluating csReq in Prolog will result in a yes or no answer. However, standard Prolog deduction cannot determine what more must be done to meet the requirements if they have not already been fulfilled; it would require analysis to find out why the deduction of csReq failed.

This sort of task can be accomplished by abductive reasoning. Given what is known in regard to which courses have been taken, what assumptions could be made to render provable the statement that all graduation requirements have been met?

This paper extends an earlier paper [18] that did not include a description of the chained specific abduction scheme and its inference rules. Chained specific abduction provides a means for propagating assumption costs from literals in the formula being proved to literals obtained by backward chaining; these inherited costs are a very useful feature for natural-language interpretation [8].

2   Four Abduction Schemes

We will consider here the abductive explanation of conjunctions of positive literals from Horn clause knowledge bases. An explanation will consist of a substitution for variables in the conjunction and a set of literals to be assumed. In short, we are developing an abductive extension of pure Prolog.

---

[1]Alternative abductive approaches to natural-language interpretation have been proposed by Charniak [3] and Norvig [12].

The general approach can be characterized as follows: when trying to explain why $Q(a)$ is true, hypothesize $P(a)$ if $P(x) \supset Q(x)$ is known.

The requirement that assumptions be literals does not permit us to explain $Q(a)$ when $P(a)$ is known by assuming $P(x) \supset Q(x)$, or even $P(a) \supset Q(a)$. We do not regard this as a limitation in tasks such as diagnosis and natural-language interpretation. Some other tasks, such as scientific-theory formation, could be cast in terms of abductive explanation when the assumptions take these more general forms.

We want to include the possibility that $Q(a)$ can be explained by assuming $Q(a)$. As later examples will show, this is vital in the natural-language interpretation task.

Consider again the example of the deductive database for graduation requirements. All the possible ways of fulfilling the requirements can be obtained by backward chaining from csReq:

```
<- csReq.
<- basicCS, mathReq, advancedCS, engReq, natSciReq.
<- basicCS, mathReq, advancedCS, engReq, physicsI, physicsII.
<- basicCS, mathReq, advancedCS, engReq, chemI, chemII.
<- basicCS, mathReq, advancedCS, engReq, bioI, bioII.
<- basicCS, mathReq, advancedCS, digSys, natSciReq.
<- basicCS, mathReq, advancedCS, digSys, physicsI, physicsII.
<- basicCS, mathReq, advancedCS, digSys, chemI, chemII.
<- basicCS, mathReq, advancedCS, digSys, bioI, bioII.
        :
```

Eliminating from any such clause those requirements that have been met results in a list that, if met, would result in fulfilling the graduation requirements. Different clauses can be more or less specific about how the remaining requirements must be satisfied. If the student lacks only Physics II to graduate, the backward-chaining scheme can derive the statements that he or she can fulfill the requirements for graduation by satisfying physicsII, natSciReq, or (rather uninformatively) csReq.

The above clauses are all possible abductive explanations for meeting the graduation requirements.

In general, if the formula $Q_1 \wedge \cdots \wedge Q_n$ is to be explained or abductively proved, the substitution [of values for variables] θ and the assumptions $P_1, \ldots, P_m$ would constitute one possible explanation if $(P_1 \wedge \cdots \wedge P_m) \supset (Q_1 \wedge \cdots \wedge Q_n)\theta$ is a consequence of the knowledge base.

If, in the foregoing example, the student lacks only Physics II to graduate, assuming physicsII then makes csReq provable.

If the explanation contains variables, such as $P(x)$ as an assumption to explain $Q(x)$, the explanation should be interpreted as neither to assume $P(x)$ for all x (i.e., assume $\forall x P(x)$) nor to assume $P(x)$ for some unspecified x (i.e., assume $\exists x P(x)$), but rather that, for any variable-free instance t of x, if $P(t)$ is assumed, then $Q(t)$ follows.

It is a general requirement that the conjunction of all assumptions made be consistent with the knowledge base. In the natural-language interpretation task, the rejection of assumptions that are inconsistent with the knowledge base presupposes that the knowledge base is correct and that the speaker of the sentence is neither mistaken nor lying.

With an added factoring operation and without the literal ordering restriction, so that any, not just the leftmost, literal of a clause can be resolved on, Prolog-style backward chaining is capable of generating all possible explanations that are consistent with the knowledge base. That is, every possible explanation consistent with the knowledge base is subsumed by an explanation that is generable by backward chaining and factoring.

It would be desirable if the procedure were guaranteed to generate no explanations that are inconsistent with the knowledge base. However, this is impossible, although fortunately not all inconsistent explanations are generated; the system can generate only those explanations that assume literals reached from the initial formula by backward chaining. Consistency of explanations with the knowledge base must be checked outside the abductive-reasoning inference system. Determining consistency is undecidable in general, though decidable subcases do exist, and many explanations can be rejected quickly for being inconsistent with the knowledge base. For example, assumptions can be readily rejected if they violate sort or ordering restrictions, e.g., assuming $woman(John)$ can be disallowed if $man(John)$ is known or already assumed, and assuming $b < a$ can be disallowed if $a \leq b$ is known or already assumed. Sort restrictions are particularly effective in eliminating inconsistent explanations in natural-language interpretation. We shall not discuss the consistency requirement further; what we are primarily concerned with here is the process of generating possible explanations, in order of preference according to our cost criteria, not with the extra task of verifying their consistency with the knowledge base.

Obviously, any clause derived by backward chaining and factoring can be used as a list of assumptions to prove the correspondingly instantiated initial formula abductively. This can result in an overwhelming number of possible explanations. Various abductive schemes have been developed to limit the number of acceptable explanations. These schemes differ in their specification of which literals are assumable.

What we shall call *most specific abduction* has been used particularly in diagnostic tasks. In explaining symptoms in a diagnostic task, the objective is to identify causes that, if assumed to exist, would result in the symptoms. The most specific causes are usually sought, since identifying less specific causes may not be as useful. In most specific abduction, the only literals that can be assumed are those to which backward chaining can no longer be applied.

What we shall call *predicate specific abduction* has been used particularly in planning and design synthesis tasks. In generating a plan or design by specifying its objectives and ascertaining what assumptions must be made to make the objectives provable, acceptable assumptions are often expressed in terms of a prespecified set of predicates. In planning, for example, these might represent the set of executable actions.

We consider what we will call *least specific abduction* to be well suited to natural-language-interpretation tasks. It allows only literals in the initial formula to be assumed.

Given that abductive reasoning has been used mostly for diagnosis and planning, and that least specific abduction tends to produce what would be considered frivolous results for such tasks, least specific abduction has been little studied. Least specific abduction is used in natural-language interpretation to seek the least specific assumptions that explain a sentence. More specific explanations would unnecessarily and often incorrectly require excessively detailed assumptions.

Although least specific abduction is often sufficient for natural-language interpretation, it is clearly sometimes necessary to assume literals that are not in the initial formula. We propose *chained specific abduction* for these situations. Assumability is inherited—a literal can be assumed if it is an assumable literal in the initial formula or if it can be obtained by backward chaining from an assumable literal.

2.1 Most Specific Abduction

Resolution based systems for abductive reasoning applied to diagnostic tasks [13,4,5] have favored the most specific explanations by adopting as assumptions only pure literals, which cannot be resolved with any clause in the knowledge base, that are reached by backward chaining from the formula to be explained. For causal-reasoning tasks, this eliminates frivolous and unhelpful explanations for "the watch is broken" such as simply noting that the watch is broken, as opposed to, perhaps, noting the mainspring is broken. Also, explanations can be too specific. In diagnosing the failure of a computer system, most specific abduction could never merely report the failure of a board if the knowledge base has enough information about the board structure for the failure to be explained, possibly in many inconsistent ways, by the failure of its components.

Besides sometimes providing overly specific explanations, as discussed further in Section 2.3, the pure-literal based most specific abduction scheme is incomplete: it does not compute all the reasonable most specific explanations.

Consider explaining instances of the formula $P(x) \land Q(x)$ with a knowledge base that consists of $P(a)$ and $Q(b)$. For most specific abduction, backward chaining to sets of pure literals makes $P(c) \land Q(c)$ explainable by assuming $P(c)$ and $Q(c)$ as both literals are pure, but $P(x) \land Q(x)$ is explainable only by assuming $P(b)$ or $Q(a)$, since $P(x)$ and $Q(x)$ are not pure. The explanation will not be found that assumes $P(c)$ and $Q(c)$, or any value of x other than a or b, to explain $P(x) \land Q(x)$.

Thus, most specific abduction does not lift properly from the case of variable-free formulas to the general case; this would not be a problem if we restricted ourselves to propositional calculus formulas. A solution in the general case would be to require that all generalizations of any pure literal also be pure. However, this is often impractical, since the purity of $P(c)$ in the above example would require the purity of $P(x)$, which is inconsistent with the presence of $P(a)$ in the knowledge base.

A special case of the requirement that generalizations of pure literals be pure would be to have a set of predicates that do not occur positively, i.e., they appear only in negated literals, in the knowledge base. But the case of a set of assumable predicate symbols is handled more generally, without the purity requirement, by predicate specific abduction (see Section 2.2). This is consistent with much of the practice in diagnostic tasks, where causal explanations in terms of particular predicates, such as Ab, are often sought.

2.2 Predicate Specific Abduction

Resolution based systems for abductive reasoning applied to planning and design synthesis tasks [6] have favored explanations expressed in terms of a prespecified subset of the predicates, namely, the assumable predicates.

In explaining $P(x) \wedge Q(x)$ with a knowledge base that consists of $P(a)$ and $Q(b)$, predicate specific abduction would offer the following explanations: (1) $Q(b)$, if P is assumable, (2) $P(a)$, if Q is assumable, along with (3) $P(x) \wedge Q(x)$, if both are assumable.

2.3 Least Specific Abduction

The criterion for "best explanation" used in natural-language interpretation differs greatly from that used in most specific abduction for diagnostic tasks. To interpret the sentence "the watch is broken," the conclusion will likely be that we should add to our knowledge the information that the watch currently discussed is broken. The explanation that would be frivolous and unhelpful in a diagnostic task is just right for sentence interpretation. A more specific causal explanation, such as a broken mainspring, would be gratuitous.

Associating the assumability of a literal with its purity, as most specific abduction does, yields not only causally specific explanations, but also taxonomically specific explanations. With axioms such as $mercury(x) \supset liquid(x)$ and $water(x) \supset liquid(x)$, explaining $liquid(a)$, when $liquid(a)$ cannot be proved, would require the assumption that a was mercury, or that it was water, and so on. Not only are these explanations more specific than the only fully warranted one that a is simply a liquid, but none may be correct: for example, a might be milk, but milk is not mentioned as a possible liquid. Most specific abduction thus assumes completeness of the knowledge base with respect to causes, subtypes, and so on. The purity requirement may make it impossible to make any assumption at all. Many reasonable axiom sets contain axioms that make literals, which we would sometimes like to assume, impure and unassumable. For example, in the presence of $parent(x,y) \supset child(y,x)$ and $child(x,y) \supset parent(y,x)$, neither $child(a,b)$ nor $parent(b,a)$ could be assumed, since neither literal is pure.

We note that assuming any literals, other than those in the initial formula, generally results in more specific and thus more risky assumptions. When explaining R with $P \supset R$ (or $P \wedge Q \supset R$) in the knowledge base, either R or P (or P and Q) can be assumed to explain R. Assumption of R, the consequent of an implication, in preference to the antecedent P (or P and Q), results in the fewest consequences. Assuming the antecedent may result in more consequences, e.g., if other rules such as $P \supset S$ are present.

Predicate specific abduction is not ideal for natural-language interpretation either, since there is no easy division of predicates into assumable and nonassumable, so that those assumptions that can be made will be reasonably restricted. Most predicates must be assumable in some circumstances such as when certain sentences are being interpreted, but in many other cases should not be assumed.

Least specific abduction, wherein a subset of the literals asked to be proven must be assumed, comes closer to our ideal of the right method of explanation for natural-language interpretation. Under this model, a sentence is translated into a logical form that contains literals whose predicates stand for properties and relationships and whose variable and constant arguments refer to entities specified or implied by the sentence. The logical form is then proved abductively, with some or all of the variable values filled in from the knowledge base and the unprovable literals of the logical form assumed.

The motivation for this is the claim that what we should learn from a sentence is often near the surface and can attained by assuming literals in the logical form of the sentence. For example, when interpreting the sentence

The car is red.

with the logical form

$car(x) \wedge red(x)$,[2]

we would typically want to ascertain from the discourse which car x is being discussed and learn by abductive assumption that it is red and not something more specific, such as the fact that it is carmine or belongs to a fire chief (whose cars, according to the knowledge base, might always be red).

2.4 Chained Specific Abduction

In least specific abduction, only literals in the initial formula can be assumed. Although this yields the correct result in many cases, it is clearly sometimes necessary to make deeper assumptions that imply the initial formula. When interpreting a piece of text which refers to fish and pets, with the logical form

$fish(x) \wedge pet(y) \wedge \cdots$

$fish(x)$ and $pet(y)$ must be assumed, if no fish or pets are in the knowledge base.

But we would like to consider the possibility that x and y refer to the same entity; we could do this by least specific abduction only if (in our knowledge base) all fish are pets or all pets are fish, so we could assume one and use it to prove the other.

What is needed are axioms like

$fish(x) \wedge fp(x) \supset pet(x)$ or $pet(x) \wedge pf(x) \supset fish(x)$

---

[2]A logical form that insisted upon proving $car(x)$ and assuming $red(x)$ might have been used instead. We prefer this more neutral logical form to allow for alternative interpretations. The preferred interpretation is determined by the assignment of costs to axioms and assumable literals.

which state that fish are sometimes pets or that pets are sometimes fish. The predicates fp and pf denote the extra requirements for a fish to be a pet or a pet to be a fish.

Effective use of such axioms requires that literals other than those in the initial formula be assumable. When backward chaining with an implication, chained specific abduction allows the antecedent literals of the implication to inherit assumability from the literal that matches the consequent of the implication.

Because $pet(y)$ is assumable, backward-chained to literals $fish(y)$ and $fp(y)$ may be assumable. Either $fish(x)$ or $fish(y)$ can be assumed and used to factor the other with the result that $x = y$, and $fp(y)$ can be assumed to produce an explanation in which x and y refer to the same entity.

Factoring some literals obtained by backward chaining and assuming the remaining antecedent literals can also sometimes yield better explanations. When $Q \wedge R$ is explained from

$$P_1 \wedge P_2 \supset Q$$
$$P_2 \wedge P_3 \supset R$$

the explanation that assumes P_1, P_2, and P_3 may be preferable to the one that assumes Q and R. Even if Q and R are not provable, it might not be necessary to assume all of P_1, P_2, and P_3, since some may be provable.

3   Assumption Costs

A key issue in abductive reasoning is picking the best explanation. Defining this is so subjective and task dependent that there is no hope of devising an algorithm that will always compute only the best explanation. Nevertheless, there are often so many abductive explanations that it is necessary to have some means of eliminating most of them. We attach numeric assumption costs to assumable literals, and compute minimum-cost abductive explanations in an effort to influence the abductive reasoning system toward favoring the intended explanations.

We regard the assignment of numeric costs as a part of programming the explanation task. The values used may be determined by subjective estimates of the likelihood of various interpretations, or perhaps they may be learned through exposure to a large set of examples.

In selecting the best abductive explanation, we often prefer, given the choice, that certain literals be assumed rather than others. For example, for the sentence

The car is red.

with the logical form

$$car(x) \wedge red(x)$$

the knowledge base will likely contain both cars and things that are red. However, the form of the sentence suggests that $red(x)$ is new information to be learned and that $car(x)$ should be proved from the knowledge base because it is derived from a definite reference, i.e., a specific car is presumably being discussed. Thus, an explanation that assumes $red(a)$ where $car(a)$ is provable should be preferred to an explanation that assumes $car(b)$ where $red(b)$ is provable. A way to express this preference is through the assumption costs associated with the literals: $car(x)$ could have cost 10, and $red(x)$ cost 1.

The cost of an abductive explanation could then be the sum of the assumption costs of all the literals that had to be assumed: $car(a) \wedge red(a)$ would be the preferred explanation, with cost 1, and $car(b) \wedge red(b)$ would be another explanation, with the higher cost 10.

However, if only the cost of assuming literals is counted in the cost of an explanation, there is in general no effective procedure for computing a minimum-cost explanation. For example, if we are to explain P, where P is assumable with cost 10, then assuming P produces an explanation with cost 10, but proving P would result in a better explanation with cost 0. Since provability of first-order formulas is undecidable in general, it may be impossible to determine whether the cost 10 explanation is best.

The solution to this difficulty is that the cost of proving literals, as well as the cost of assuming them, must be included in the cost of an explanation. An explanation that assumes P with cost 10 would be preferred to an explanation that proves P with cost 50 (e.g., in a proof of 50 steps) but would be rejected in favor of an explanation that proves P with cost less than 10.

Treating explanation costs as composed only of assumption costs is attractive: why should we distinguish explanations that differ in the size of their proof, when only their provability should matter? However, there are substantial advantages gained by taking into account proof costs as well as assumption costs, in addition to the crucial benefit of making theoretically possible the search for a minimum-cost explanation.

If costs are associated with the axioms in the knowledge base as well as with assumable literals, these costs can be used to encode information on the likely relevance of the fact or rule to the situation in which the sentence is being interpreted.

Axiom costs can be adjusted to reflect the salience of certain facts. If a is a car mentioned in the previous sentence, the cost of the axiom $car(a)$ could be adjusted downward so that the explanation of $car(x) \wedge red(x)$ that assumes $red(a)$ would be preferred to one that assumes $red(c)$ for some other car c in the knowledge base.

Indeed, the explanation that assumes $red(a)$ should probably be preferred to any explanation that proves both $car(c)$ and $red(c)$, i.e., there is a red car c in the knowledge base, even though this last would be a perfect zero-cost explanation if only assumption costs were used, because the recent mention of a makes it likely that a is the subject of the sentence, and the purpose of the sentence is to convey the new information that a car is red. Interpreting the referent of "the car" as a car that is already known to be red results in no new information being learned.

We have some reservations about choosing explanations on the basis of numeric costs. Nonnumeric specification of preferences is an important research topic. Nevertheless, we have found these numeric costs to be quite practical; they offer an easy way of specifying that one literal is to be assumed rather than another. When many alternative explanations are possible, summing numeric costs in each explanation, and adopting an explanation with minimum total cost, provides a mechanism for comparing the costs of one proof and set of assumptions against the costs of another. If this method of choosing explanations is too simple, other means may be too complex to be realizable, since they would require preference choices among a wide variety of sets of assumptions and proofs. We provide a procedure for computing a minimum-cost explanation by enumerating possible partial explanations in order of increasing cost. Even a perfect scheme for specifying preferences among alternative explanations may not lead to an effective procedure for generating a most preferred one, as there may be no way of cutting off the search with the certainty that the best explanation exists among those so far discovered. Finally, any scheme will be imperfect: people may disagree as to the best explanation of some data and, moreover, sometimes do misinterpret sentences.

4 Minimum-Cost Proofs

We now present the inference system for computing abductive explanations. This method applies to predicate specific, least specific, and chained specific abduction. We have not tried to incorporate most specific abduction into this scheme because of its incompleteness, its incompatibility with ordering restrictions, and its unsuitability for natural-language interpretation.

Every literal Q_i in the initial formula is annotated with its assumption cost c_i:

$$Q_1^{c_1}, \ldots, Q_n^{c_n}$$

The cost c_i must be nonnegative; it can be infinite, if Q_i is not to be assumed.

Every literal P_j in the antecedent of an implication in the knowledge base is annotated with its assumability function f_j:

$$P_1^{f_1}, \ldots, P_m^{f_m} \supset Q$$

The input and output values for each f_i are nonnegative and possibly infinite. If this implication is used to backward chain from $Q_i^{c_i}$, then the literals $P_1, \ldots, P_m$ will be in the resulting formula with assumption costs $f_1(c_i), \ldots, f_m(c_i)$.

In predicate specific abduction, costs are associated with predicates, so assumptions costs are the same for all occurrences of the predicate. Let $cost(p)$ denote the assumption cost for predicate p. The assumption cost c_i for literal Q_i in the initial formula is $cost(p)$, where the Q_i predicate is p; the assumption function f_j for literal P_j in the antecedent of an implication is the unary function whose value is uniformly $cost(p)$, where the P_j predicate is p.

In least specific abduction, different occurrences of the predicate in the initial formula may have different assumption costs, but only literals in the initial formula are assumable.

The assumption cost c_i for literal Q_i in the initial formula is arbitrarily specified; the assumption function f_j for literal P_j in the antecedent of an implication has value infinity.

In chained specific abduction, the most general case, different occurrences of the predicate in the initial formula may have different assumption costs; literals obtained by backward chaining can have flexibly computed assumption costs that depend on the assumption cost of the literal backward-chained from. The assumption cost c_i for literal Q_i in the initial formula is arbitrarily specified; the assumption function f_j for literal P_j in the antecedent of an implication can be an arbitrary monotonic unary function.

We have most often used simple weighting functions of the form $f_j(c) = w_j \times c\,(w_j > 0)$. Thus, the implication

$$P_1^{w_1} \wedge P_2^{w_2} \supset Q$$

states that P_1 and P_2 imply Q, but also that, if Q is assumable with cost c, then P_1 is assumable with cost $w_1 \times c$ and P_2 is assumable with cost $w_2 \times c$, as the result of backward chaining from Q. If $w_1 + w_2 < 1$, more specific explanations are favored, since the cost of assuming P_1 and P_2 is less than the cost of assuming Q. If $w_1 + w_2 > 1$, less specific explanations are favored: Q will be assumed in preference to P_1 and P_2. But, depending on the weights, P_i might be assumed in preference to Q if P_j is provable.

The cost of a proof is usually taken to be a measure of the syntactic form of the proof, e.g., the number of steps in the proof. A more abstract characterization of cost is needed. We want to assign different costs to different inferences by associating costs with individual axioms; we also want to have a cost measure that is not so dependent on the syntactic form of the proof.

We assign to each axiom A a cost $axiom\text{-}cost(A)$ that is greater than zero. Assumption costs $assumption\text{-}cost(L)$ are computed for each literal L. When viewed abstractly, a proof is a demonstration that the goal follows from a set S of substitution instances of the axioms, together with, in the case of abductive proofs, a set H of literals that are assumed in the proof. We want to count the cost of each separate instance of an axiom or assumption only once instead of the number of times it may appear in the syntactic form of the proof. Thus, a natural measure of the cost of the proof is

$$\sum_{A\sigma \in S} axiom\text{-}cost(A) + \sum_{L \in H} assumption\text{-}cost(L)$$

Consider the example of explaining $Q(x) \wedge R(x) \wedge S(x)$ with a knowledge base that includes $P(a)$, $P(x) \supset Q(x)$, and $Q(x) \wedge R(x) \supset S(x)$, and with R assumable. By using Prolog plus an inference rule for assuming literals, we get:

```
1. <- Q(x), R(x), S(x).
2. <- P(x), R(x), S(x).      % resolve 1 with Q(x) <- P(x)
3. <- R(a), S(a).            % resolve 2 with P(a)
4. <- S(a).                  % assume R(a) in 3
5. <- Q(a), R(a).            % resolve 4 with S(x) <- Q(x), R(x)
```

```
6. <- P(a), R(a).          % resolve 5 with Q(x) <- P(x)
7. <- R(a)                 % resolve 6 with P(a)
8. <- true                 % assume R(a) in 7
```

$Q(x) \land R(x) \land S(x)$ is explained with x having the value a under the assumption that $R(a)$ is true.

The cost of the proof is the sum of the costs of the axiom instances $P(a)$, $P(a) \supset Q(a)$, and $Q(a) \land R(a) \supset S(a)$, plus the cost of assuming $R(a)$. The costs of using $P(a)$ and $P(x) \supset Q(x)$ and assuming $R(a)$ are not counted twice even though they were used twice, since the same instances were used or assumed. If, however, we had used $P(x) \supset Q(x)$ with b as well as a substituted for x, then the cost of $P(x) \land Q(x)$ would have been counted twice.

In general, the cost of a proof can be determined by extracting the sets of axiom instances S and assumptions H from the proof tree and performing the above computation. However, it is an enormous convenience if there always exists a *simple proof tree* such that each separate instance of an axiom or assumption actually occurs only once in the proof tree. That way, as the inferences are performed, costs can simply be added to compute the cost of the current partial proof. Even if the same instance of an axiom or assumption happens to be used and counted twice, a different, cheaper derivation would use and count it only once. Partial proofs can be enumerated in order of increasing cost by employing breadth-first or iterative-deepening search methods and minimum-cost explanations can be discovered effectively. Iterative-deepening search is compatible with maintaining Prolog-style implementation and performance [17,19,20].

We shall describe our inference system as an extension of pure Prolog. Prolog, though complete for Horn sets of clauses, lacks this desirable property of always being able to yield a simple proof tree.

Prolog's inference system—ordered input resolution without factoring—would have to eliminate the ordering restriction and add the factoring operation to remain a form of resolution and be able to prove Q, R from $Q \leftarrow P$, $R \leftarrow P$, and P without using P twice. Elimination of the ordering restriction is potentially very expensive. For example, there are $n!$ proofs of $Q_1, \ldots, Q_n$ from the axioms $Q_1, \ldots, Q_n$ when unordered input resolution is used, but only one with ordered input resolution. Implementations of most specific abduction perform unordered input resolution [13,4,5].

We present a resolution-like inference system, an extension of pure Prolog, that preserves the ordering restriction and does not require repeated use of the same instances of axioms. In our extension, literals in goals can be marked with information that dictates how the literals are to be treated by the inference system, whereas in Prolog, all literals in goals are treated alike and must be proved. A literal can be marked as one of the following:

proved The literal has been proved or is in the process of being proved; in this inference system, a literal marked as proved will have been fully proved when no literal to its left remains unsolved.

assumed The literal is being assumed.

unsolved The literal is neither proved nor assumed.

The initial goal clause $Q_1, \ldots, Q_n$ in a deduction consists of literals Q_i that are either unsolved or assumed. If any assumed literals are present, they must precede the unsolved literals. Unsolved literals must be proved from the knowledge base plus any assumptions in the initial goal clause or made during the proof, or, in the case of assumable literals, may be directly assumed. Literals that are proved or assumed are retained in all successor goal clauses in the deduction and are used to eliminate matching goals. The final goal clause $P_1, \ldots, P_m$ in a deduction must consist entirely of proved or assumed literals P_i.

An abductive proof is a sequence of goal clauses $G_1, \ldots, G_p$ for which

- G_1 is the initial goal clause.

- each G_{k+1} $(1 \leq k < p)$ is derived from G_k by resolution with a fact or rule, making an assumption, or factoring with a proved or assumed literal.

- G_p has no unsolved literals (all are proved or assumed).

These rules differ substantially from those presented in our earlier paper [18], which were sufficient for predicate specific and least specific abduction, but not for chained specific abduction.

Predicate specific abduction is quite simple because the assumability and assumption cost of a literal are determined by its predicate symbol. Least specific abduction is also comparatively simple because if a literal is not provable or assumable and must be factored, all assumable literals with which it can be factored are present in the initial and derived formulas. Because assumability is inherited in chained specific abduction, the absence of a literal to factor with is not a cause for failure. Such a literal may appear in a later derived clause after further inference as new, possibly assumable, literals are introduced by backward chaining.

4.1 Inference Rules

Suppose the current goal G_k is $Q_1^{c_1}, \ldots, Q_n^{c_n}$ and that $Q_i^{c_i}$ is the leftmost unsolved literal. Then the following inferences are possible.

4.1.1 Resolution with a fact

Let axiom A be a fact Q with its variables renamed, if necessary, so that it has no variables in common with the goal G_k. Then, if Q_i and Q are unifiable with most general unifier σ, the goal

$$G_{k+1} = Q_1^{c_1}\sigma, \ldots, Q_n^{c_n}\sigma$$

with

$$cost'(G_{k+1}) = cost'(G_k) + axiom\text{-}cost(A)$$

can be derived, where $Q_i\sigma$ is marked as proved in G_{k+1}.[3]

The resolution with a fact or rule operations differ from their Prolog counterparts principally in the retention of $Q_i\sigma$ (marked as proved) in the result. Its retention allows its use in future factoring.

4.1.2 Resolution with a rule

Let axiom A be a rule $Q \leftarrow P_1^{f_1}, \ldots, P_m^{f_m}$ with its variables renamed, if necessary, so that it has no variables in common with the goal G_k. Then, if Q_i and Q are unifiable with most general unifier σ, the goal

$$G_{k+1} = Q_1^{c_1}\sigma, \ldots, Q_{i-1}^{c_{i-1}}\sigma, P_1^{f_1(c_i)}\sigma, \ldots, P_m^{f_m(c_i)}\sigma, Q_i^{c_i}\sigma, \ldots, Q_n^{c_n}\sigma$$

with

$$cost'(G_{k+1}) = cost'(G_k) + axiom\text{-}cost(A)$$

can be derived, where $Q_i\sigma$ is marked as proved in G_{k+1} and each $P_j\sigma$ is unsolved.

4.1.3 Making an assumption

The goal

$$G_{k+1} = G_k$$

with

$$cost'(G_{k+1}) = cost'(G_k)$$

can be derived, where Q_i is marked as assumed in G_{k+1}.

Similarly to resolution, Q_i is retained in the result, for use in future factoring.

The same result, except for Q_i being marked as proved instead of assumed, could be derived by resolution with a fact if assumable literals are asserted as axioms. The final proof could then be examined to distinguish between proved and assumed literals. Although using a fact and making an assumption can be merged operationally in this way, we prefer to regard them as separate operations. An important distinction between facts and assumable literals is that facts are consistent with the assumed-consistent knowledge base; assumptions made in an abductive explanation should be checked for consistency with the knowledge base before being accepted.

---

[3] Each literal in a goal G_{k+1} resulting from one of these inference rules is proved or assumed precisely when its parent literal in G_k is, unless it is stated otherwise.

4.1.4 Factoring with a proved or assumed literal

If Q_i and Q_j $(j < i)^4$ are unifiable with most general unifier σ, the goal

$$G_{k+1} = Q_1^{c_1}\sigma, \ldots, Q_{j-1}^{c_{j-1}}\sigma, Q_j^{c_j'}\sigma, Q_{j+1}^{c_{j+1}}\sigma, \ldots, Q_{i-1}^{c_{i-1}}\sigma, Q_{i+1}^{c_{i+1}}\sigma, \ldots, Q_n^{c_n}\sigma$$

with

$$cost'(G_{k+1}) = cost'(G_k)$$

can be derived, where $c_j' = min(c_j, c_i)$.

Note that if Q_j is a proved literal and $c_j' < c_j$, the assumption costs of assumed literals descended from Q_j may need to be adjusted also. Thus, in resolution with a rule, it may be necessary to retain assumption costs $f_1(c_i), \ldots, f_m(c_i)$ in symbolic rather than numeric form, so that they can be readily updated if a later factoring operation changes the value of c_i.

4.1.5 Computing Cost of Completed Proof

If no literal of G_k is unsolved (all are proved or assumed) and $Q_{i_1}, \ldots, Q_{i_m}$ are the assumed literals of G_k,

$$cost(G_k) = cost'(G_k) + \sum_{i \in \{i_1, \ldots, i_m\}} c_i$$

Consider again the example of explaining $Q(x) \wedge R(x) \wedge S(x)$ with R assumable from a knowledge base that includes $P(a)$, $P(x) \supset Q(x)$, and $Q(x) \wedge R(x) \supset S(x)$. Proved literals are marked by brackets [], assumed literals by braces {}.

```
1. <- Q(x), R(x), S(x).
2. <- P(x), [Q(x)], R(x), S(x).        % resolve 1 with Q(x) <- P(x)
3. <- [P(a)], [Q(a)], R(a), S(a).      % resolve 2 with P(a)
4. <- [P(a)], [Q(a)], {R(a)}, S(a).    % assume R(a) in 3
5. <- [P(a)], [Q(a)], {R(a)}, Q(a), R(a), [S(a)].
                                       % resolve 4 with S(x) <- Q(x), R(x)
6. <- [P(a)], [Q(a)], {R(a)}, R(a), [S(a)]. % factor 5
7. <- [P(a)], [Q(a)], {R(a)}, [S(a)]. % factor 6
```

The abductive proof is complete when all literals are either proved or assumed. Each axiom instance and assumption was used or made only once in the proof.

The proof procedure can be restricted to disallow any clause in which there are two identical proved or assumed literals. Identical literals should have been factored if neither

---

[4] Q_j must have been proved or assumed, since it precedes Q_i.

was an ancestor of the other. Alternative proofs are also possible whenever a literal is identical to an ancestor literal [9,10,15].

If no literals are assumed, the procedure is a disguised form of Shostak's graph construction (GC) procedure [15] restricted to Horn clauses, where proved literals play the role of Shostak's C-literals. It also resembles Finger's ordered residue procedure [6], except that the latter retains assumed literals (rotating them to the end of the clause) but not proved literals. Thus, it includes the ability of the GC procedure to compute simple proof trees for Horn clauses and the ability of the ordered residue procedure to make assumptions in abductive proofs.

Another approach which shares the idea of using least cost proofs to choose explanations is Post's Least Exception Logic [14]. This is restricted to the propositional calculus, with first-order problems handled by creating ground instances, because it relies upon a translation of default reasoning problems into integer linear programming problems. It finds sets of assumptions, defined by default rules, that are sufficient to prove the theorem, that are consistent with the knowledge base so far as it has been instantiated, and that have least cost.

4.2 Search Strategy Refinements

Unless the axioms are carefully written to preclude infinite branches in the search space, the standard unbounded depth-first search strategy of Prolog is inadequate. Because of the possibility of making assumptions, branches are even less likely to be terminated by failure than in regular Prolog processing. Thus, we have generally executed this inference system with depth-first iterative deepening search with $cost'$ bounded.

The value of $cost'$ is incremented by the resolution rules, but not by the assumption or factoring rules. Factoring does not increase the cost of the final proof, so it is correct for $cost'$ to be not incremented in that case. Making an assumption will generally increase the cost of the proof, but the amount is uncertain when the assumption is made, since the assumed literal might later be factored with another literal with a lower assumption cost. Because the final assumption cost, after such factoring, may be zero, $cost'$ is incremented by zero so that $cost'$ remains an admissable, never overestimating, estimator of the final proof cost $cost$, and iterative-deepening search will be guaranteed to find proofs in order of increasing cost.

If assumption operations do not increment $cost'$, then assumptions can be made and proofs found that are immediately rejected as too costly when the cost of the completed proof is computed. An extreme case often occurs when assuming a literal whose assumption cost is infinite; assuming such a literal will lead to an infinite cost proof, unless the literal is factored with another literal with finite assumption cost. These zero-cost assumption operations can result in large search space.

This problem can be mitigated in a number of ways. These generally entail incrementing $cost'$ when making assumptions; this results in more search cutoffs, as the bound on $cost'$ is more often exceeded.

Assumption of literals with infinite cost can often be eliminated by creating a list of all predicates that never have finite assumption costs or functions. These literals need never be assumed, since there is no possibility of the literal being factored with another literal with finite assumption cost, and the proof cost cannot be reduced to a finite value.

A lower bound on the assumption cost can be specified on a predicate-by-predicate basis. In the case of those predicates that never have finite assumption costs or functions, the lower bound can be infinite. With this lower bound instead of the implied lower bound of zero, $cost'$ is incremented by the lower bound on assumption cost for the predicate of the assumed literal. When computing the cost of a completed proof, only the excess of the assumption costs over their lower bounds is added to $cost'$ to compute $cost$.

A more extreme approach is to simply increment $cost'$ by the assumption cost of a literal as it is assumed. ($cost'$ must be incremented by some smaller finite value in the case of those literals with infinite assumption cost that might be factorable with a literal with finite assumption cost.) The value of $cost'$ must later be decremented if the literal is factored with another literal with a lower assumption cost. Because under these conditions $cost'$ may sometimes overestimate the final proof cost, this results in an inadmissable search strategy: proofs cannot be guaranteed to be found in order of increasing cost. Nevertheless, this approach may work well in practice, if factoring with a literal with significantly lower assumption cost is infrequent enough.

5 Future Directions

A valuable extension of this work would be to allow for non-Horn sets of axioms.

Computing minimum-cost proofs from non-Horn sets of axioms is more difficult and would take us farther from Prolog-like inference systems. A mutually resolving set of clauses is a set of clauses such that each clause can be resolved with every other. Shostak [16] proved that mutually resolving sets of clauses, with no tautologies and with no single atom occurring in every clause, do not have simple proof trees. This result is true of the GC procedure as well as of resolution. So, although we were able to use the GC procedure to compute simple proof trees for sets of Horn clauses, this cannot be done for non-Horn sets.

For non-Horn clause proofs, an assumption mechanism can be added to a resolution based inference system that is complete for non-Horn clauses such as the GC procedure or the model elimination procedure that is implemented in PTTP [17,19], with more complicated rules for counting costs to compensate for the absence of simple proof trees.

Alternatively, an assumption mechanism can be added to the matings or connection method [1,2]. These proof procedures do not require multiple occurrences of the same instances of axioms. This approach would reduce requirements on the syntactic form of the axioms (e.g., the need for clauses) so that a cost could be associated with an arbitrary axiom formula instead of a clause. It would be useful to allow axioms of the form $P_1 \wedge P_2 \supset Q \wedge R$, so that the axiom need be used and cost added only once in proving $Q \wedge R$. The rationale is, if P_1 and P_2 are proved or assumed in order to abductively prove Q, R should also be provable at no additional cost.

6 Conclusion

We have formulated part of the natural-language-interpretation task as abductive inference. The process of interpreting sentences in discourse can be viewed as the abductive inference of those assumptions to be made for the listener to know that the sentence is true. The forms of abduction suggested for diagnosis, and for design synthesis and planning, are generally unsuitable for natural-language interpretation. We suggest that least specific abduction, in which only literals in the logical form can be assumed, is useful for natural-language interpretation. Chained specific abduction generalizes least specific abduction to allow literals obtained by backward chaining to be assumed as necessary.

Numeric costs can be assigned to axioms and assumable literals so that the intended interpretation of a sentence will hopefully be obtained by computing the minimum-cost abductive explanation of the sentence's logical form. Axioms can be assigned different costs to reflect their relevance to the sentence. Different literals in the logical form can be assigned different assumption costs according to the form of the sentence, with literals from indefinite references being more readily assumable than those from definite references. In chained specific abduction, assumability functions can be associated with literals in the antecedents of implications, to very flexibly specify at what cost literals obtained by backward chaining can be assumed.

We have presented a Prolog-like inference system that computes abductive explanations by means of either predicate specific or least specific abduction. The inference system is designed to compute the cost of an explanation correctly, so that multiple occurrences of the same instance of an axiom or assumption are not charged for more than once.

Most of the ideas presented here have been implemented in the TACITUS project for text understanding at SRI [7,8].

Acknowledgements

Jerry Hobbs has been extremely helpful and supportive in the development of these abduction schemes for natural-language interpretation and their implementation and use in the TACITUS project. Douglas Appelt has been the principal direct user of implementations of abduction in the TACITUS system; writing axioms and assigning assumption costs and weights, he has suggested a number of enhancements to control the search space. This work has been greatly facilitated by discussions with them and Douglas Edwards, Todd Davies, John Lowrance, and Mabry Tyson.

References

[1] Andrews, P.B. Theorem proving via general matings. *Journal of the ACM 28*, 2 (April 1981), 193–214.

[2] Bibel, W. *Automated Theorem Proving*. Friedr. Vieweg & Sohn, Braunschweig, West Germany, 1982.

[3] Charniak, E. Motivation analysis, abductive unification, and nonmonotonic equality. *Artificial Intelligence 34*, 3 (April 1988), 275–295.

[4] Cox, P.T. and T. Pietrzykowski. Causes for events: their computation and applications. *Proceedings of the 8th Conference on Automated Deduction*, Oxford, England, July 1986, 608–621.

[5] Cox, P.T. and T. Pietrzykowski. General diagnosis by abductive inference. *Proceedings of the 1987 Symposium on Logic Programming*, San Francisco, California, August 1987, 183–189.

[6] Finger, J.J. *Exploiting Constraints in Design Synthesis*. Ph.D. dissertation, Department of Computer Science, Stanford University, Stanford, California, February 1987.

[7] Hobbs, J.R. and P. Martin. Local pragmatics. *Proceedings of the Tenth International Conference on Artificial Intelligence*, Milan, Italy, August 1987, 520–523.

[8] Hobbs, J.R., M. Stickel, P. Martin, and D. Edwards. Interpretation as abduction. *Proceedings of the 26th Annual Meeting of the Association for Computational Linguistics*, Buffalo, New York, June 1988, 95–103.

[9] Loveland, D.W. A simplified format for the model elimination procedure. *Journal of the ACM 16*, 3 (July 1969), 349–363.

[10] Loveland, D.W. *Automated Theorem Proving: A Logical Basis*. North-Holland, Amsterdam, the Netherlands, 1978.

[11] Maier, D. and D.S. Warren. *Computing with Logic*. Benjamin/Cummings, Menlo Park, California, 1988.

[12] Norvig, P. Inference in text understanding. *Proceedings of the AAAI-87 Sixth National Conference on Artificial Intelligence*, Seattle, Washington, July 1987, 561–565.

[13] Pople, H.E.,Jr. On the mechanization of abductive logic. *Proceedings of the Third International Joint Conference on Artificial Intelligence*, Stanford, California, August 1973, 147–152.

[14] Post, S.D. Default reasoning through integer linear programming. Planning Research Corporation, McLean, Virginia, 1988.

[15] Shostak, R.E. Refutation graphs. *Artificial Intelligence 7*, 1 (Spring 1976), 51–64.

[16] Shostak, R.E. On the complexity of resolution derivations. Unpublished, 1976(?).

[17] Stickel, M.E. A Prolog technology theorem prover: implementation by an extended Prolog compiler. *Journal of Automated Reasoning 4*, 4 (December 1988), 353–380.

[18] Stickel, M.E. A Prolog-like inference system for computing minimum-cost abductive explanations in natural-language interpretation. *Proceedings of the International Computer Science Conference '88*, Hong Kong, December 1988, 343–350.

[19] Stickel, M.E. A Prolog technology theorem prover: a new exposition and implementation in Prolog. Technical Note 464, Artificial Intelligence Center, SRI International, Menlo Park, California, June 1989.

[20] Stickel, M.E. and W.M. Tyson. An analysis of consecutively bounded depth-first search with applications in automated deduction. *Proceedings of the Ninth International Joint Conference on Artificial Intelligence*, Los Angeles, California, August 1985, 1073–1075.

Lecture Notes in Computer Science